Precalculus
A Self-Teaching Guide

Precalculus
A Self-Teaching Guide

Steve Slavin
Ginny Crisonino

John Wiley & Sons, Inc.
New York • Chichester • Weinheim • Brisbane • Singapore • Toronto

This book is printed on acid-free paper. ∞

Copyright © 2001 by Steve Slavin and Ginny Crisonino. All rights reserved
Published by John Wiley & Sons, Inc.
Published simultaneously in Canada

No part of this publication may be reproduced, stored in a retrieval system, or transmitted in any form or by any means, electronic, mechanical, photocopying, recording, scanning, or otherwise, except as permitted under Section 107 or 108 of the 1976 United States Copyright Act, without either the prior written permission of the Publisher, or authorization through payment of the appropriate per-copy fee to the Copyright Clearance Center, 222 Rosewood Drive, Danvers, MA 01923, (978) 750-8400, fax (978) 750-4744. Requests to the Publisher for permission should be addressed to the Permissions Department, John Wiley & Sons, Inc., 605 Third Avenue, New York, NY 10158-0012, (212) 850-6011, fax (212) 850-6008, email: PERMREQ@WILEY.COM.

This publication is designed to provide accurate and authoritative information in regard to the subject matter covered. It is sold with the understanding that the publisher is not engaged in rendering professional services. If professional advice or other expert assistance is required, the services of a competent professional person should be sought.

ISBN 978-1-62045-621-7

Contents

1 The Basics — 1
    Pretest, 2
    1 Exponents, 5
    2 Polynomials, 10
    3 Rational Exponents and Radicals, 16
    4 Factoring, 17
    5 Basic Geometry, 24
    6 Applications, 28

2 Functions — 33
    1 Definition of a Function, 33
    2 Operations on Functions, 41
    3 Composite Functions, 43
    4 Inverse Functions, 45

3 Graphs of Functions — 52
    1 Intercepts, 53
    2 Slope of a Straight Line, 57
    3 Writing the Equation of a Straight Line, 59
    4 Graphs of Linear Functions, 62
    5 Graphs of Quadratic Functions, 67
    6 Graphs of Polynomial Functions of Degree 3 and Higher, 76

7 Asymptotes, 82
8 Oblique Asymptotes and Graphs of Rational Functions, 88

4 Exponential and Logarithmic Functions — 103
1 Exponential Functions, 103
2 Logarithmic Functions, 109
3 Properties of Logarithmic Functions, 111
4 Solving Logarithmic Equations, 114
5 Applications, 119

5 Trigonometry — 127
1 Angles and Their Measure, 127
2 Right-Triangle Trigonometry, 132
3 Trigonometric Functions of Any Angle, 134
4 Graphs of the Basic Trigonometric Functions, 140
5 Inverse Trigonometric Functions, 142
6 Applications, 146

6 Analytic Trigonometry — 152
1 Using Fundamental Identities, 152
2 Verifying Trigonometric Identities, 156
3 Solving Trigonometric Equations, 159
4 Sum and Difference Formulas, 164
5 Multiple-Angle and Product-to-Sum Formulas, 169

7 Additional Topics in Trigonometry — 178
1 Law of Sines, 178
2 Law of Cosines, 187
3 Area of a Triangle, 191

8 Miscellaneous Topics — 196
1 Solving Systems of Inequalities and Linear Programming, 196
2 Partial Fraction Decomposition, 204

Index — 211

Precalculus
A Self-Teaching Guide

1 The Basics

Before proceeding to precalculus, we'll review some basic concepts. This chapter provides a bridge from elementary algebra to intermediate algebra and other topics in precalculus. Most of these concepts should be somewhat familiar, and you should be able to quickly pick up where you left off in algebra, whether it was last week, last year, or even the last millennium.

When you've completed this chapter you should be able to work with:

- exponents
- polynomials
- rational exponents and radicals
- factoring
- basic geometry
- applications

Before we begin, try the following pretest to get an idea of how much review you need, if any. If you get a perfect score on the pretest, you can skip chapter 1 and go directly to chapter 2. To learn precalculus you have to have your intermediate algebra down cold.

PRECALCULUS

PRETEST

Simplify the following expressions. Do not leave negative exponents in your final answer. Leave all answers in fully reduced form.

Part 1: Exponents
Simplify the following:

1. $y^3 y^4$
2. $(y^3)^4$
3. $(3a^4)^2$
4. 2^0
5. 4^{-2}
6. $\left(\dfrac{1}{4}\right)^{-2}$
7. $-2x^{-4}$
8. $(-2x)^{-4}$
9. $\dfrac{4r^2 s^{-1}}{2r^2 s^{-2}}$
10. $\left(\dfrac{3x^2 y^{-1}}{x^{-1} y^2}\right)^{-2}$

Part 2: Polynomials

11. $2x + 3y - 4x + 5y$
12. $(5a - 3b)^2$
13. $7a^3(4a^2 - 5a) - 2a^2(3a^3 - 6a^2)$
14. $(x + y)^3$
15. $7a^3(4a^2 - 5a) - 2a^2(3a^3)(-6a^2)$
16. $(2a + b)^2 - (2a - b)^2$
17. $(7u^2 v^2 + 4uv^2 - 5u) - (4u^2 v^2 - 3uv^2 + 5)$
18. $(9a^2 b + 12a^2 b^2 + 15ab) \div 3b$
19. $(8y^3 - 18y^2 - 6 + 11y) \div (-3y + 2 + 4y^2)$

Part 3: Factoring
Factor the following:

20. $15a - 3b + 12c$
21. $10b^3 c^2 - 5b^2 c$
22. $x^2 + 8x + 16$
23. $x^2 - 7x - 18$
24. $3a^2 - 5ab - 12b^2$
25. $x^2 - 100$
26. $4x^4 - 16$
27. $8x^3 - 27$
28. $y^6 + 1$
29. $2ax + bx + 2ay + by$

Part 4: Rational Exponents and Radicals
Simplify the following:

30. $8^{\frac{2}{3}}$
31. $4^{-\frac{1}{2}}$
32. $x^{\frac{5}{2}}$
33. $\left(\dfrac{9}{8}\right)^{\frac{3}{2}}$
34. $\left(\dfrac{8}{27}\right)^{-\frac{2}{3}}$
35. $\sqrt{\dfrac{32x^3}{9x}}$
36. $\sqrt[4]{32x^5}$
37. $\dfrac{2 - \sqrt{5}}{2 + 3\sqrt{5}}$
38. $2\sqrt{2x - 1} = 8$

Part 5: Basic Geometry

39. Find the length of the unknown side in the triangle below.

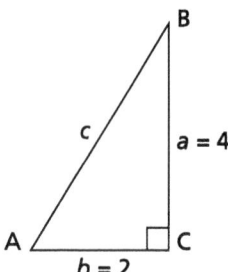

40. Find the circumference and the area of a circle with diameter 10 inches.

41. Find the volume of the rectangular solid with length 10, height 2, and width 4.

42. Find the distance of the line segment below and its midpoint.

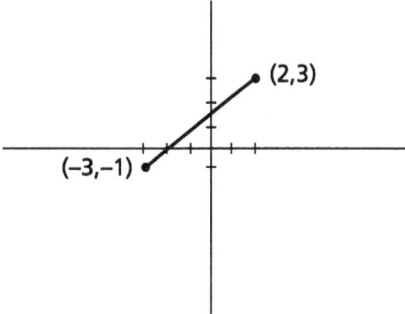

Part 6: Applications

43. A specialty coffee shop wishes to blend a $6-per-pound coffee with an $11-per-pound coffee to produce a blend that will sell for $8 per pound. How much of each should be used to produce 300 pounds for the new blend?

44. An investor has $20,000 to invest. If part grows at an interest rate of 8 percent and the rest at 12 percent, how much should be invested at each rate to yield 11 percent on the total amount invested?

45. A boat takes 1.5 times as long to go 360 miles up a river than to return. If the boat cruises at 15 miles per hour in still water, what is the rate of the current?

PRECALCULUS

ANSWERS:

1. y^7 **2.** y^{12} **3.** $9a^8$ **4.** 1

5. $\dfrac{1}{4^2} = \dfrac{1}{16}$ **6.** $(4)^2 = 16$ **7.** $-\dfrac{2}{x^4}$ **8.** $\dfrac{1}{(-2x)^4} = \dfrac{1}{16x^4}$

9. $\dfrac{4r^2s^2}{2r^2s} = 2s$ **10.** $\left(\dfrac{3x^2x}{y^2y}\right)^{-2} = \left(\dfrac{3x^3}{y^3}\right)^{-2} = \left(\dfrac{y^3}{3x^3}\right)^2 = \dfrac{y^6}{9x^6}$ **11.** $-2x + 8y$

12. $(5a - 3b)(5a - 3b)$
$25a^2 - 15ab - 15ab + 9b^2$
$25a^2 - 30ab + 9b^2$

13. $28a^5 - 35a^4 - 6a^5 + 12a^4 = 22a^5 - 23a^4$

14. $(x + y)(x + y)(x + y)$
$(x^2 + xy + xy + y^2)(x + y)$
$(x^2 + 2xy + y^2)(x + y)$
$x^3 + x^2y + 2x^2y + 2xy^2 + xy^2 + y^3$
$x^3 + 3x^2y + 3xy^2 + y^3$

15. $28a^5 - 35a^4 + 36a^7$

16. $(2a + b)(2a + b) - (2a - b)(2a - b)$
$(4a^2 + 2ab + 2ab + b^2) - (4a^2 - 2ab - 2ab + b^2)$
$4a^2 + 4ab + b^2 - 4a^2 + 4ab - b^2$
$8ab$

17. $7u^2v^2 + 4uv^2 - 5u - 4u^2v^2 + 3uv^2 - 5 = 3u^2v^2 + 7uv^2 - 5u - 5$

18. $\dfrac{9a^2b}{3b} + \dfrac{12a^2b^2}{3b} + \dfrac{15ab}{3b}$
$3a^2 + 4a^2b + 5a$

19. $4y^2 - 3y + 2 \overline{\smash{\big)}\, 8y^3 - 18y^2 + 11y - 6}$ quotient $2y - 3$
$\underline{8y^3 - 6y^2 + 4y}$
$-12y^2 + 7y - 6$
$\underline{-12y^2 + 9y - 6}$
$-2y$

$2y - 3 - \dfrac{2y}{4y^2 - 3y + 2}$

20. $3(5a - b + 4c)$ **21.** $5b^2c(2bc - 1)$ **22.** $(x + 4)^2$

23. $(x - 9)(x + 2)$ **24.** $(3a + 4b)(a - 3b)$ **25.** $(x - 10)(x + 10)$

26. $4(x^4 - 4) = 4(x^2 - 2)(x^2 + 2)$ **27.** $(2x - 3)(4x^2 + 6x + 9)$ **28.** $(y^2 + 1)(y^4 - y^2 + 1)$

29. $2ax + bx + 2ay + by$
$x(2a + b) + y(2a + b)$
$(2a + b)(x + y)$

30. $(\sqrt[3]{8})^2 = (2)^2 = 4$

31. $\dfrac{1}{4^{\frac{1}{2}}} = \dfrac{1}{\sqrt{4}} = \dfrac{1}{2}$ **32.** $\sqrt{x^5} = x^2\sqrt{x}$ **33.** $\left(\sqrt{\dfrac{9}{8}}\right)^3 = \left(\dfrac{3}{2\sqrt{2}}\right)^3 = \dfrac{27}{8(2\sqrt{2})} \cdot \dfrac{\sqrt{2}}{\sqrt{2}}$
$= \dfrac{27\sqrt{2}}{16(2)} = \dfrac{27\sqrt{2}}{32}$

34. $\dfrac{1}{\left(\sqrt[3]{\dfrac{8}{27}}\right)^2} = \dfrac{1}{\left(\dfrac{2}{3}\right)^2} = \dfrac{1}{\dfrac{4}{9}} = \dfrac{9}{4}$ **35.** $\sqrt{\dfrac{32x^2}{9}} = \dfrac{4x\sqrt{2}}{3}$ **36.** $2x\sqrt[4]{2x}$

37. $\left(\dfrac{2-\sqrt{5}}{2+3\sqrt{5}}\right)\left(\dfrac{2-3\sqrt{5}}{2-3\sqrt{5}}\right) = \dfrac{4-6\sqrt{5}-2\sqrt{5}+15}{-41} = \dfrac{-19+8\sqrt{5}}{41}$

38. $2\sqrt{2x-1} = 8$
$(\sqrt{2x-1})^2 = 4^2$
$2x - 1 = 16$
$2x = 17$
$x = \dfrac{17}{2}$

39. $a^2 + b^2 = c^2$
$4^2 + 2^2 = c^2$
$16 + 4 = c^2$
$20 = c^2$
$c = 2\sqrt{5}$

40. $C = 2\pi(5) = 10\pi$
$A = \pi(5)^2 = 25\pi$

41. $v = l(w)(h) = 10(4)(2) = 80$

42. $(2,3); (-3,-1)$
$d = \sqrt{(x_2 - x_1)^2 + (y_2 - y_1)^2}$
$d = \sqrt{(-3-2)^2 + (-1-3)^2}$
$d = \sqrt{25 + 16}$
$d = \sqrt{41}$

$md = \left(\dfrac{x_1 + x_2}{2}, \dfrac{y_1 + y_2}{2}\right)$

$md = \left(\dfrac{2-3}{2}, \dfrac{3-1}{2}\right)$

$md = \left(-\dfrac{1}{2}, 1\right)$

43. $6x + 11(300 - x) = 8(300)$
$6x + 3300 - 11x = 2400$
$-5x = -900$
$x = 180$ lbs. of \$6 coffee
120 lbs. of \$11 coffee

44. $.08x + .12(20000 - x) = .11(20000)$
$8x + 12(20000 - x) = 11(20000)$
$8x + 240000 - 12x = 220000$
$-4x = -20000$
$x = 5000$
\$5,000 at 8%, \$15,000 at 12%

45. $\dfrac{360}{15 - x} = \dfrac{1.5(360)}{15 + x}$
$5400 + 360x = 8100 - 540x$
$x = 3$ miles per hour

1 Exponents

As you should already be aware, exponents are a way of showing repeated multiplication. For example, 2^3 (read "two cubed" or "two to the third power") means we multiply $2(2)(2) = 8$. The 2 is the base and the 3 is the exponent, or power. The exponent tells us how many times we use the base as a factor. So, $a^3 = a(a)(a)$. The reverse is also true: we could say that $x(x)(x)(x)(x) = x^5$ (read "x to the fifth power"). There are nine basic laws of exponents you should know for precalculus. As we go over these laws, we'll ask you to try some, then check your answers with ours. There will be problems where you will have to apply more than one of these laws. Generally you don't have to apply them in the same order we have. There is no preset order of laws used for simplifying exponents. Let's get started.

Law 1:

$x^a x^b = x^{a+b}$ When multiplying, if the bases are the same, add the exponents.

Example 1:
$x^3 x^5 = x^8$

Example 2:
$y^2 y^7 y = y^{10}$ Don't forget y is $1y^1$.

Example 3:
$2^3 2^2 = 2^5 = 32$

Example 4:
$(2a^3 bc^5)(3ab^2 c) = 6a^4 b^3 c^6$ Multiply the coefficients, add the exponents.

Law 2:

$(x^a)^b = x^{ab}$ When raising a power to a power, multiply the exponents.

Example 5:
$(x^3)^5 = x^{15}$

Example 6:
$(y^2)^7 = y^{14}$

Example 7:
$(2^3)^2 = 2^6 = 64$

Law 3:

$(cx^a)^b = c^b x^{ab}$ The c is a constant. If a constant is in parentheses with an exponent on the outside of the parentheses, the constant is also raised to that power.

Example 8:
$(2x^5)^3 = 2^3 x^{15} = 8x^{15}$

Example 9:
$(5y^7 z^{11})^4 = 5^4 y^{28} z^{44} = 625 y^{28} z^{44}$

Law 4:

$$\left(\frac{x^a}{y^b}\right)^n = \frac{x^{an}}{y^{bn}}$$

When raising a fraction to a power, multiply the exponents in both the numerator and the denominator by the power.

Example 10:

$$\left(\frac{s^3}{t^5}\right)^7 = \frac{s^{21}}{t^{35}}$$

Example 11:

$$\left(\frac{u^2}{v^4}\right)^3 = \frac{u^6}{v^{12}}$$

Before we go over the remaining laws of exponents, try the following problems:

SELF-TEST 1:

1. $e^2 e^3$ 2. $(4d^3 e^8)^3$ 3. $(f^2)^4$ 4. $f^2 f^4$ 5. $(7t^2)(2t^4)$

ANSWERS:

1. e^5 2. $64d^9 e^{24}$ 3. f^8 4. f^6 5. $14t^6$

Law 5:
If $x \neq 0$

$$\frac{x^a}{x^b} = \begin{cases} x^{a-b}, & \text{if } a > b \\ \frac{1}{x^{b-a}}, & \text{if } b > a \\ 1, & \text{if } a = b \end{cases}$$

Example 12:

$$\frac{y^6}{y^2} = y^4 \qquad \frac{yyyyyy}{yy} = y^4$$

Example 13:

$$\frac{y^2}{y^6} = \frac{1}{y^4} \qquad \frac{yy}{yyyyyy} = \frac{1}{y^4}$$

Example 14:

$$\frac{y^2}{y^2} = 1 \qquad \frac{\cancel{y}\cancel{y}}{\cancel{y}\cancel{y}} = 1$$

Law 6:

$x^0 = 1$, if $x \neq 0$ 	Anything (except 0) to the zero power is 1.

Example 15:

$y^0 = 1$

Example 16:

$7^0 = 1$

Example 17:

$(4ab^{11})^0 = 1$

Law 7:

$x^{-a} = \dfrac{1}{x^a}$ 	Anything (except 0) to a negative power becomes a fraction, with 1 over itself to the positive power.

Example 18:

$y^{-4} = \dfrac{1}{y^4}$

Example 19:

$2t^{-5} = \dfrac{2}{t^5}$ 	Notice that only t^5 moved down to the denominator, not the 2. Because the 2 isn't in parentheses, it isn't raised to the negative fifth power. See the next example for a slightly different case.

Example 20:

$(2t)^{-5} = \dfrac{1}{(2t)^5} = \dfrac{1}{32t^5}$ 	Since both the 2 and the t are inside the parentheses, they are both raised to the negative fifth power, so they both move to the denominator.

The Basics 9

Law 8:

$$\frac{1}{x^{-a}} = \frac{1}{\frac{1}{x^a}} = x^a$$

When a base in the denominator is raised to a negative power, the base moves up to the numerator and the exponent changes to a positive.

Example 21:

$$\frac{4}{x^{-4}} = 4x^4$$

The x^{-4} in the denominator moved up to the numerator as x^4.

Example 22:

$$\frac{s^{-3}t^3}{s^2t^{-2}} = \frac{t^2t^3}{s^2s^3} = \frac{t^5}{s^5}$$

The s^{-3} moved down to the denominator as an s^3, and the t^{-2} moved up to the numerator as t^2.

Law 9:

$$\left(\frac{x}{y}\right)^{-a} = \left(\frac{y}{x}\right)^a$$

A fraction to a negative power becomes its reciprocal to the positive power. Exponents inside the parentheses do not change sign when you flip the fraction.

Example 23:

$$\left(\frac{s}{t}\right)^{-3} = \left(\frac{t}{s}\right)^3 = \frac{t^3}{s^3}$$

Example 24:

$$\left(\frac{b^3}{c^2}\right)^{-4} = \left(\frac{c^2}{b^3}\right)^4 = \frac{c^8}{b^{12}}$$

Notice only the fraction on the outside of the parentheses changes sign when law 9 is applied.

SELF-TEST 2:

1. $\dfrac{y^2z^3}{y^4z^2}$ 2. $\dfrac{c^7}{c^7}$ 3. m^0 4. $\left(\dfrac{2}{z^3}\right)^{-2}$ 5. 3^{-2}

6. t^{-3} 7. $\dfrac{x^{-2}}{x^{-3}}$ 8. $\left(\dfrac{3t^{-1}u^2}{2t^{-2}u^0}\right)^{-4}$ 9. $\dfrac{1}{a^{-2}}$ 10. $\left(\dfrac{3x^{-1}y^2}{x^2y^{-2}}\right)^{-3}$

ANSWERS:

1. $\dfrac{z}{y^2}$ 2. 1 3. 1 4. $\left(\dfrac{z^3}{2}\right)^2 = \dfrac{z^6}{4}$ 5. $\dfrac{1}{3^2} = \dfrac{1}{9}$ 6. $\dfrac{1}{t^3}$

7. $\dfrac{x^{-2}}{x^{-3}} = \dfrac{x^3}{x^2} = x$ Notice that if we move the negative exponents from the numerator to the denominator, or from the denominator to the numerator, the exponents become positive.

8. $\left(\dfrac{3t^{-1}u^2}{2t^{-2}u^0}\right)^{-4} = \left(\dfrac{2t^{-2}}{3t^{-1}u^2}\right)^4 = \left(\dfrac{2^4t^{-8}}{3^4t^{-4}u^8}\right) = \dfrac{16t^4}{81t^8u^8} = \dfrac{16}{81t^4u^8}$ 9. a^2

10. $\left(\dfrac{3x^{-1}y^2}{x^2y^{-2}}\right)^{-3} = \left(\dfrac{x^2y^{-2}}{3x^{-1}y^2}\right)^3 = \dfrac{x^6y^{-6}}{3^3x^{-3}y^6} = \dfrac{x^6x^3}{3^3y^6y^6} = \dfrac{x^9}{27y^{12}}$

2 Polynomials

A polynomial is an algebraic expression involving addition, subtraction, or multiplication of terms whose exponents are non-negative integers. That means no negative or fractional exponents. That also means no variables in any denominator. Some examples of polynomials are:

$2x + 4y \qquad a^3b - 2xyz^5 \qquad 5s^3t^2 - 10st^3 \qquad \dfrac{3r^3s}{5} + 4rs^{10}$

Some examples of algebraic expressions that are *not* polynomials are:

$\dfrac{3r}{s}$ This is not a polynomial because there's a variable in the denominator.

$4u^{-2} + 8uv$ This is not a polynomial because there's a negative exponent.

$3x^{\frac{1}{2}} - 6xy^3$ This is not a polynomial because there's a fractional exponent.

When we have studied polynomials more we'll find that they are very useful when we want to write an algebraic statement to represent a real-life situation. We'll see some examples of this in chapter 2, when we study functions.

Now that we know what a polynomial is, it's time for us to add and subtract polynomials. When we add or subtract the terms of polynomials, it's called combining like terms. Remember, combining like terms applies only to addition and subtraction. Let's look first at what like terms are.

Like terms are the same variables to the same powers. Notice that our definition doesn't require like terms to have the same coefficients, or to be in any particular order.

Some like terms are:

$3x^2$ and $6x^2$ $4a^3b^6$ and $-5b^6a^3$ $8r^2s^4t^7$ and $3r^2s^4t^7$

Some algebraic expressions that are *not* like terms are:

$4x^2$ and $6x$ $30a^3b^6$ and $-5a^6b^3$ $4r^3s^4t^7$ and $4r^7s^4t^3$

Combining like terms: When we combine like terms we add the coefficients, not the exponents. For example, if we wanted to add $2x^3$ and $4x^3$ we'd have $6x^3$, not $6x^6$.

Example 25:
$5x^7 + 4y^5 - 2x^7 - 8y^5 = 3x^7 - 4y^5$

Example 26:
$3a^3b^5 + 2a^2b + a^2b - 4ab^2 = 3a^3b^5 + 3a^2b - 4ab^2$

Example 27:
$3st^2 + 5u^4vw^{21} - st^2 - 6u^4vw^{21} - st^2 + u^4vw^{21} = st^2$

SELF-TEST 3:
1. $3z^4 + 5z^4$
2. $5xyz^9 - 6xyz^9$
3. $abc + abc + abc + abc$
4. $2st^6 - 6s^2t^8 + 6s^2t^8 - 2st^6$
5. $10ed^3 - 12ed^3 + 15e^3d^4 - 11e^3d^4 - 2ed^3$

ANSWERS:
1. $8z^4$
2. $-xyz^9$
3. $4abc$
4. 0
5. $-4ed^3 + 4e^3d^4$

Now that we've reviewed combining like terms, it's time for us to multiply polynomials. When we multiply terms, we multiply the coefficients and add the exponents. For example, when we multiply $4x^3$ times $3x^5 + 5x^2$ we multiply the coefficients and add the exponents. The $4x^3$ "distributes" over both the $3x^5$ and the $5x^2$.

$4x^3(3x^5 + 5x^2) = 12x^8 + 20x^5$

Example 28:

$-(5u^2 + 3v^3 - 4w) = -1(5u^2 + 3v^3 - 4w) = -5u^2 - 3v^3 + 4w$

Whenever a – sign without a number is in front of the parentheses, we assume it's a –1. A – sign in front of a grouping **reverses the sign** of every term inside the grouping.

Example 29:

$(7st^3 - 3s^2t)(4s^2t - st^3)$
$28s^3t^4 - 7s^2t^6 - 12s^4t^2 + 3s^3t^4$
$31s^3t^4 - 7s^2t^6 - 12s^4t^2$

Distribute the terms in the first set of parentheses over the terms in the second set of parentheses, then combine the like terms.

Example 30:

$(4a^2b + 3ab)(ab + 3a^2b - a^3b)$

$4a^3b^2 + 12a^4b^2 - 4a^5b^2 + 3a^2b^2 + 9a^3b^2 - 3a^4b^2$
$13a^3b^2 + 9a^4b^2 - 4a^5b^2 + 3a^2b^2$

Example 31:

$(s^3 - 5t^4)^2$
$(s^3 - 5t^4)(s^3 - 5t^4)$
$s^6 - 5s^3t^4 - 5s^3t^4 + 25t^8$
$s^6 - 10s^3t^4 + 25t^8$

Be careful: $(s^3 - 5t^4)^2 \neq (s^3)^2 - (5t^4)^2$. This is a very common error; watch out! You cannot square the terms inside the parentheses; first you must write the binomial twice and distribute. This example is the square of a difference, not the difference of squares.

SELF-TEST 4:

1. $3x^4(4x + 5b^2)$
2. $6a^2b(2a^3 - 3ab^2 + 4ab^2)$
3. $-2(x + y - z) - (3x + 4y - 5z)$
4. $(3s^3 + 2t^4)^2$
5. $(4x^2 + 2x)(3x^2 - x)$
6. $(x - y)^2 - (x + y)^2$
7. $(a - b^8)^3$

ANSWERS:

1. $12x^5 + 15x^4b^2$

2. $12a^5b - 18a^3b^3 + 24a^3b^3$
   $12a^5b + 6a^3b^3$

3. $-2x - 2y + 2z - 3x - 4y + 5z$
   $-5x - 6y + 7z$

4. $(3s^3 + 2t^4)(3s^3 + 2t^4)$
   $9s^6 + 6s^3t^4 + 6s^3t^4 + 4t^8$
   $9s^6 + 12s^3t^4 + 4t^8$

5. $12x^4 - 4x^3 + 6x^3 - 2x^2$
   $12x^4 + 2x^3 - 2x^2$

6. $(x - y)(x - y) - (x + y)(x + y)$
   $x^2 - xy - xy + y^2 - (x^2 + xy + xy + y^2)$
   $x^2 - 2xy + y^2 - (x^2 + 2xy + y^2)$
   $x^2 - 2xy + y^2 - x^2 - 2xy - y^2$
   $-4xy$

7. $(a - b^8)(a - b^8)(a - b^8)$
$(a^2 - ab^8 - ab^8 + b^{16})(a - b^8)$
$(a^2 - 2ab^8 + b^{16})(a - b^8)$
$a^3 - a^2b^8 - 2a^2b^8 + 2ab^{16} + ab^{16} - b^{24}$
$a^3 - 3a^2b^8 + 3ab^{16} - b^{24}$

Now that we've reviewed addition, subtraction, and multiplication of polynomials, it's time for division of polynomials. There are two types of division involving polynomials: division by a monomial (one term) and by a binomial (two terms). Let's start with division by a monomial. To divide a polynomial by a monomial, we put each term of the polynomial over the divisor and reduce the fractions. For example, suppose we wanted to divide $(4m^2 + 6m + 10)$ by 2. We would put each term over 2 and reduce the fractions.

$$\frac{4m^2}{2} + \frac{6m}{2} + \frac{10}{2} = 2m^2 + 3m + 5$$

Example 32:

Divide $(4x^3 - 16x^2 + 12x)$ by $6x^3$.

$\dfrac{4x^3}{6x^3} - \dfrac{16x^2}{6x^3} + \dfrac{12x}{6x^3}$ Since this is division by a monomial, we separate the polynomial into three fractions, then reduce.

$\dfrac{2}{3} - \dfrac{8}{3x} + \dfrac{2}{x^2}$

Example 33:

$\dfrac{9x^2y - 36xy^2}{9x^2y}$

$\dfrac{9x^2y}{9x^2y} - \dfrac{36xy^2}{9x^2y}$ Remember, anything divided by itself is 1.

$1 - \dfrac{4y}{x}$

SELF-TEST 5:

1. Divide $(4x^3 - 16x^2 + 12x)$ by $4x$.
2. $\dfrac{10a^2 + 5a - 15}{5a^2}$
3. $(4k^4 - 2k^3 + 12k^2 - 6k) \div 2k^3$
4. $\dfrac{6a^2b + 12a^2b^2 + 18ab}{6ab^2}$

5. Divide $(25s^3 + 15s^2 - 10s)$ by $5s$.

6. $\dfrac{30x^4y^2z^2 - 15x^2y^4z^2 + 20x^2y^2}{5x^2y^4z}$

ANSWERS:

1. $\dfrac{4x^3}{4x} - \dfrac{16x^2}{4x} + \dfrac{12x}{4x}$

$x^2 - 4x + 3$

2. $\dfrac{10a^2}{5a^2} + \dfrac{5a}{5a^2} - \dfrac{15}{5a^2}$

$2 + \dfrac{1}{a} - \dfrac{3}{a^2}$

3. $\dfrac{4k^4}{2k^3} - \dfrac{2k^3}{2k^3} + \dfrac{12k^2}{2k^3} - \dfrac{6k}{2k^3}$

$2k - 1 + \dfrac{6}{k} - \dfrac{3}{k^2}$

4. $\dfrac{6a^2b}{6ab^2} + \dfrac{12a^2b^2}{6ab^2} + \dfrac{18ab}{6ab^2}$

$\dfrac{a}{b} + 2a + \dfrac{3}{b}$

5. $\dfrac{25s^3}{5s} + \dfrac{15s^2}{5s} - \dfrac{10s}{5s}$

$5s^2 + 3s - 2$

6. $\dfrac{30x^4y^2z^2}{5x^2y^4z} - \dfrac{15x^2y^4z^2}{5x^2y^4z} + \dfrac{20x^2y^2}{5x^2y^4z}$

$\dfrac{6x^2z}{y^2} - 3z + \dfrac{4}{y^2z}$

Now it's time for division of a polynomial by a binomial. For this type of problem we use a division box like in arithmetic. Suppose we want to divide $(10x^4 - 6x - 15 + 5x^3)$ by $(2x + 1)$; here's how we do it.

Example 34:

$(10x^4 + 5x^3 - 6x - 15) \div (2x + 1)$

The first step in all of these problems is to put the polynomial terms in descending order. To write a polynomial in descending order we begin with the term with the highest exponent, in this case $10x^4$. Then we write the term with the next highest exponent, $5x^3$, and follow the same procedure with the other terms until all of them have been ordered. If there are any missing terms, such as the x^2 term in this polynomial, we fill it in with a $0x^2$. All terms must be accounted for. Next we use a division box and start dividing.

$$\begin{array}{r} 5x^3 - 3 - \dfrac{12}{2x+1} \\ 2x+1 \overline{\smash{)}10x^4 + 5x^3 + 0x^2 - 6x - 15} \\ \underline{10x^4 + 5x^3} \\ -6x - 15 \\ \underline{-6x - 3} \\ -12 \end{array}$$

Our first question is what's $\dfrac{10x^4}{2x}$? It's $5x^3$. Now we multiply $5x^3(2x + 1)$. Make sure you line up like terms under like terms, then subtract. What's $\frac{-6x}{2x}$? It's -3.

Multiply $-3(2x + 1)$ and subtract. Remember, we're subtracting. $-15 - (-3) = -15 + 3 = -12$. Our remainder goes over the divisor.

Example 35:

$(7x^6 + 1 + x - 3x^2 - 14x^4) \div (x^2 - 2)$ Write the polynomial in descending order.

$$x^2 - 2 \overline{\smash{)}7x^6 + 0x^5 - 14x^4 + 0x^3 - 3x^2 + x + 1}$$

with quotient $7x^4 \quad\quad\quad -3 + \dfrac{x-5}{x^2-2}$

$\dfrac{7x^6}{x^2} = 7x^4$, multiply $7x^4(x^2 - 2)$.

$7x^6 \quad\quad -14x^4$

Be sure to line up like terms.
$-14x^2 - (-14x^4) = -14x^2 + 14x^2 = 0$.

$-3x^2 + x + 1$
$-3x^2 \quad\quad + 6$
$\quad\quad\quad x - 5$

$\dfrac{-3x^2}{x^2} = -3$

SELF-TEST 6:

1. Divide $(15x^3 + 4x^2 - 15x + 4)$ by $(3x - 1)$.
2. $(y + 7y^4 - 3y^2 + 2y^6 - 7y^5) \div (2y^2 - y)$.
3. $(x^4 - 1) \div (x^2 - 1)$.
4. Divide $(8a^5 + 2a^4 - 12a^3 + a^2 + a + 2)$ by $(4a + 1)$.

ANSWERS:

1.
$$3x - 1 \overline{\smash{)}15x^3 + 4x^2 - 15x + 4}$$
quotient: $5x^2 + 3x - 4$

$15x^3 - 5x^2$
$\quad\quad 9x^2 - 15x + 4$
$\quad\quad 9x^2 - 3x$
$\quad\quad\quad\quad -12x + 4$
$\quad\quad\quad\quad -12x + 4$
$\quad\quad\quad\quad\quad\quad 0$

Watch out for the double negatives when you subtract!

$4x^2 - (-5x^2) = 4x^2 + 5x^2 = 9x^2$

$-15x - (-3x) = -15x + 3x = -12x$

2.
$$2y^2 - y \overline{\smash{)}2y^6 - 7y^5 + 7y^4 + 0y^3 - 3y^2 + y}$$
quotient: $y^4 - 3y^3 + 2y^2 + y - 1$

$2y^6 - y^5$
$\quad\quad -6y^5 + 7y^4$
$\quad\quad -6y^5 + 3y^4$
$\quad\quad\quad\quad 4y^4 + 0y^3$
$\quad\quad\quad\quad 4y^4 - 2y^3$
$\quad\quad\quad\quad\quad\quad 2y^3 - 3y^2$
$\quad\quad\quad\quad\quad\quad 2y^3 - y^2$
$\quad\quad\quad\quad\quad\quad\quad\quad -2y^2 + y$
$\quad\quad\quad\quad\quad\quad\quad\quad -2y^2 + y$
$\quad\quad\quad\quad\quad\quad\quad\quad\quad\quad 0$

Make sure all terms are in descending order before you divide. Watch out for the double negatives!
$-7y^5 - (-y^5) = -7y^5 + y^5 = -6y^5$.

$0y^3 - (-2y^3) = 0y^3 + 2y^3 = 2y^3$

$-3y^2 - (-y^2) = -3y^2 + y^2 = -2y^2$

3.
$$\require{enclose}\begin{array}{r}x^2+1\\x^2-1\enclose{longdiv}{x^4+0x^3+0x^2+0x-1}\end{array}$$

$$\begin{array}{r}\underline{x^4-x^2}\\x^2-1\\\underline{x^2-1}\\0\end{array}$$

Be careful to line up like terms under like terms. Watch out for the double negatives!
$0x^2 - (-x^2) = 0x^2 + x^2 = x^2$

4.
$$4a+1\enclose{longdiv}{8a^5+2a^4-12a^3+a^2+a+2}$$

quotient: $2a^4 - 3a^2 + a$

$\underline{8a^5 + 2a^4}$
$-12a^3 + a^2$
$\underline{-12a^3 - 3a^2}$
$4a^2 + a$
$\underline{4a^2 + a}$
2

$$2a^4 - 3a^2 + a + \frac{2}{4a+1}$$

Watch out for the double negatives!

$a^2 - (-3a^2) = a^2 + 3a^2 = 4a^2$

3 Rational Exponents and Radicals

In section 1 we worked on exponents; we'll start this section by using radicals, or roots, to reverse the operation we used for exponents. Then we'll review how to convert from radical form to exponential form and from exponential form to radical form. From arithmetic we know that $3^2 = 9$. We also know that the square root of 9, written $\sqrt{9}$, equals 3, so we can see that the square root is the *inverse* of squaring. When we write $\sqrt{9}$, it's understood we mean $\sqrt[2]{9}$. In exponential form we would write this as $9^{\frac{1}{2}}$. The denominator of the exponent is the root (square root in this case), and the numerator is the exponent. Here's another case: $4^3 = 64$, $\sqrt[3]{64} = 4$. This would be written as $64^{\frac{1}{3}} = 4$.

	Exponential Form	Radical Form	Answer
Example 36:	$27^{\frac{2}{3}}$	$(\sqrt[3]{27})^2$	$(3)^2 = 9$
Example 37:	$27^{-\frac{2}{3}}$	$\dfrac{1}{(\sqrt[3]{27})^2}$	$\dfrac{1}{(3)^2} = \dfrac{1}{9}$

Let's look at some problems involving variables. For example, $(x^2)^3 = x^6$ and $\sqrt[3]{x^6} = x^2$ (notice all we had to do here was divide the exponent under the radical 6 by the root of the radical 3). Another example: $\sqrt[4]{16x^4y^8} = 2xy^2$. Here we just divided the exponents by the root 4. What happens when the exponent is not divisible by the root, such as in the problem $\sqrt[5]{32x^6y^8z^{10}}$? The fifth root of 64 is 2. We'll rewrite $\sqrt[5]{x^6}$ as $\sqrt[5]{x^5x}$, which is $x\sqrt[5]{x}$. We'll do the same with $\sqrt[5]{y^8} = \sqrt[5]{y^5y^3} = y\sqrt[5]{y^3}$, $\sqrt[5]{z^{10}} = z^2$. Now we have our answer, $\sqrt[5]{32x^6y^8z^{10}} = 2xyz^2\sqrt[5]{xy^3}$.

Example 38:

Simplify $\left(\dfrac{8}{27}\right)^{-\frac{2}{3}}$

Solution:

$\left(\dfrac{27}{8}\right)^{\frac{2}{3}} = \left(\sqrt[3]{\dfrac{27}{8}}\right)^2 = \left(\dfrac{3}{2}\right)^2 = \dfrac{9}{4}$

SELF-TEST 7:

1. $\sqrt[4]{81x^4}$
2. $(27)^{\frac{1}{3}}$
3. $(27)^{-\frac{1}{3}}$
4. $\left(-\dfrac{1}{32}\right)^{-\frac{4}{5}}$
5. $\sqrt{50a^3b^4}$
6. $\sqrt[4]{x^{60}}$
7. $\sqrt{28x^9y^{13}}$
8. $36^{-\frac{1}{2}}$
9. $\left(\dfrac{64}{27}\right)^{\frac{1}{3}}$
10. $\left(\dfrac{64}{27}\right)^{-\frac{2}{3}}$

ANSWERS:

1. $3x$
2. $\sqrt[3]{27} = 3$
3. $\dfrac{1}{\sqrt[3]{27}} = \dfrac{1}{3}$
4. $(-32)^{\frac{4}{5}} = (\sqrt[5]{-32})^4 = (-2)^4 = 16$
5. $\sqrt{25(2)a^2ab^4} = 5ab^2\sqrt{2a}$
6. x^{15}
7. $\sqrt{7(4)x^8xy^{12}y} = 2x^4y^6\sqrt{7xy}$
8. $\dfrac{1}{36^{\frac{1}{2}}} = \dfrac{1}{\sqrt{36}} = \dfrac{1}{6}$
9. $\sqrt[3]{\dfrac{64}{27}} = \dfrac{4}{3}$
10. $\left(\dfrac{27}{64}\right)^{\frac{2}{3}} = \left(\sqrt[3]{\dfrac{27}{64}}\right)^2 = \left(\dfrac{3}{4}\right)^2 = \dfrac{9}{16}$

4 Factoring

In section 2 we multiplied polynomials, which always resulted in an answer that is in addition or subtraction form. For example, $(2x + 3)(5x + 2) = 10x^2 + 19x + 6$. Now we'll reverse this process. We'll start with a polynomial in addition or subtraction form, such as $10x^2 + 19x + 6$, and we'll rewrite it as a product of its factors $(2x + 3)(5x + 2)$. When we rewrite a sum or a difference as a product, it's called factoring. In this section we'll review the basic techniques of factoring. Factoring is a very important topic that will help us in later chapters to simplify functions. Throughout this section remember our motto: *Factoring is fun!*

Technique 1: The Greatest Common Factor (G.C.F.)

This is always the first step on all factoring problems. We always look for the greatest common factor of *every* term in the polynomial before we apply any of the other factoring techniques.

Example 39:

Factor $2x^4 + 4x^3 - 16x^2$.

Solution:

What's the greatest factor common to all terms? The greatest common factor for 2, 4, and 16 is 2. The greatest common factor for x^4, x^3, and x^2 is x^2 (it's always the lowest exponent). The G.C.F. of the whole polynomial is $2x^2$. For us to factor out $2x^2$ from every term, we have to divide every term by the G.C.F. $2x^2$ to find the remaining terms of the factored form of the polynomial.

$$\frac{2x^4}{2x^2} = x^2 \quad \frac{4x^3}{2x^2} = 2x \quad \frac{-16x^2}{2x} = -8$$

$$2x^4 + 4x^3 - 16x^2 = 2x^2(x^2 + 2x - 8)$$

Example 40:

Factor $9a^4b^2 + 3a^3b$.

Solution:

First we ask ourselves "What's the G.C.F. of 9 and 3?" It's 3, of course. What's the G.C.F. of a^4 and a^3? It's a^3. What's the G.C.F. of b^2 and b? It's b. So the G.C.F. of $9a^4b^2 + 3a^3b$ is $3a^3b$. Next we factor $3a^3b$ out of each term of our polynomial.

$$\frac{9a^4b^2}{3a^3b} = 3ab \quad \frac{3a^3b}{3a^3b} = 1 \quad 9a^4b^2 + 3a^3b = 3a^3b(3ab + 1)$$

Example 41:

Factor $2a(x + 1) + 3b(x + 1)$.

Solution:

The G.C.F. this time is the binomial $(x + 1)$. Usually people expect a G.C.F. to be a monomial, but it doesn't have to be. In this case it's a binomial.

$$\frac{2a(x+1)}{(x+1)} = 2a \quad \frac{3b(x+1)}{(x+1)} = 3b \quad 2a(x+1) + 3b(x+1) = (x+1)(2a+3b)$$

SELF-TEST 8:

Factor the following:

1. $9g - 3h$
2. $10j - 5k + 30n$
3. $m^2 - 3m^3$
4. $12a^2b^2 + 2a^3b$
5. $12c^3d^4 + 3cd^3 - 3cd^2$
6. $4x(y-1) + 5z(y-1)$
7. $3s(4c-7) - 5t(4c-7)$
8. $6g(2-a) - 5h(2-a)$

ANSWERS:

1. $3(3g - h)$
2. $5(2j - k + 6n)$
3. $m^2(1 - 3m)$
4. $2a^2b(6b + a)$
5. $3cd^2(4c^2d^2 + d - 1)$
6. $(y - 1)(4x + 5z)$
7. $(4c - 7)(3s - 5t)$
8. $(2 - a)(6g - 5h)$

Technique 2: By Grouping

The second technique of factoring is called "by grouping." The clue that a factoring question may be a by grouping problem is the presence of four terms. Remember, the first step on all factoring problems is to factor out the G.C.F. Now let's have some fun factoring.

Example 42:

$2ax + 2bx + 2ay + 2by$	
$2(ax + bx + ay + by)$	Our first step is to factor out the G.C.F. of 2. We have four terms left, which is our clue that this is probably a by grouping problem. Our next step is to group together any two pairs of two terms that have a common factor.
$2[(ax + bx) + (ay + by)]$	
$2[x(a + b) + y(a + b)]$	Now we factor out the G.C.F. of each pair.
$2[(a + b)(x + y)]$	Last, we factor out the G.C.F. of $(a + b)$, and we're finished.

Example 43:

$xy + 2x + 4y + 8$	We look for a G.C.F. and find none. We'll group together any two terms that have a common factor. It doesn't matter which two are grouped together as long as they have a common factor.
$(xy + 2x) + (4y + 8)$	
$x(y + 2) + 4(y + 2)$	Factor the common factor out of each pair.
$(y + 2)(x + 4)$	Finally, factor out the G.C.F. of $(y + 2)$, and we're done.

Example 44:

$xy + 2x - 4y - 8$
$(xy + 2x) + (-4y - 8)$
$x(y + 2) - 4(y + 2)$
$(y + 2)(x - 4)$

This problem is almost the same as in example 43. The only difference is the negative sign in front of the 4y and the 8. Let's see how this will affect our factoring. This time we factored out a −4 instead of a positive 4. If we had factored out a positive 4 we would have gotten $x(y + 2) + 4(-y - 2)$, which would have prevented us from factoring out the $(y + 2)$.

Now it's your turn to try some by groupings.

SELF-TEST 9:

Factor the following:

1. $sr + 3s + 2rt + 6t$
2. $3ux + 3vx + 3uy + 3vy$
3. $3ux + 3vx - 3uy - 3vy$
4. $6 - 2w - 3z + wz$
5. $5ac + 5ad + 6bc + 6bd$
6. $8pr - 8ps + 2qr - 2qs$

ANSWERS:

1. $(sr + 3s) + (2rt + 6t)$
$s(r + 3) + 2t(r + 3)$
$(r + 3)(s + 2t)$

2. $3(ux + vx + uy + vy)$
$3[(ux + vx) + (uy + vy)]$
$3[x(u + v) + y(u + v)]$
$3[(x + y)(u + v)]$

3. $3(ux + vx - uy - vy)$
$3[(ux + vx) + (-uy - vy)]$
$3[x(u + v) - y(u + v)]$
$3[(u + v)(x - y)]$

4. $(6 - 2w) + (-3z + wz)$
$2(3 - w) + -z(3 - w)$
$(3 - w)(2 - z)$

5. $(5ac + 5ad) + (6bc + 6bd)$
$5a(c + d) + 6b(c + d)$
$(c + d)(5a + 6b)$

6. $2(4pr - 4ps + qr - qs)$
$2[(4pr - 4ps) + (qr - qs)]$
$2[(4p(r - s) + q(r - s)]$
$2[(r - s)(4p + q)]$

Technique 3: The Difference of Two Squares

$x^2 - y^2 = (x - y)(x + y)$ or $(x + y)(x - y)$

This technique applies only to differences, not to sums. When we factor a difference of two perfect squares we get two binomials, which are the same except that one will have a + sign and one will have a − sign. In section 2 we multiplied polynomials. If we were to multiply $(x + 1)(x - 1)$ we would get $x^2 - x + x - 1$, which would be $x^2 - 1$. Notice that the binomials we multiplied together are the same except for the signs; one is a + sign and one is a − sign. Their product yielded $x^2 - 1$, which is the difference of two squares. The factored form of $x^2 - 1$ is $(x - 1)(x + 1)$, or we could also say $(x + 1)(x - 1)$. The order in which we write the factored form doesn't matter, because as we already know, multiplication is commutative. That means $(x - 1)(x + 1)$ is the same as $(x + 1)(x - 1)$.

Example 45:
Factor $y^2 - 4$.

$(y + 2)(y - 2)$ Our first step is to look for a G.C.F., but there isn't one, so we can apply the difference of two squares. The square root of y^2 is y, and the square root of 4 is 2.

Example 46:
Factor $2t^6 - 18$.

$2(t^6 - 9)$
$2(t^3 - 3)(t^3 + 3)$ First we'll factor out the G.C.F. of 2. The square root of t^6 is t^3. The square root of 9 is 3.

Example 47:
Factor $x^8 - 16$.

$(x^4 + 4)(x^4 - 4)$
$(x^4 + 4)(x^2 - 2)(x^2 + 2)$ There is no G.C.F. The square root of x^8 is x^4. The square root of 16 is 4. For this problem we apply the difference of two squares twice. The second time is to $x^4 - 4$. We always factor every problem fully.

SELF-TEST 10:

Factor the following:

1. $y^2 - 25$
2. $2w^2 - 72$
3. $3s^2 - 27$
4. $u^{10} - 100$
5. $-x^2 + 49$
6. $z^{12} - 81$

ANSWERS:

1. $(y - 5)(y + 5)$
2. $2(w^2 - 36)$
   $2[(w + 6)(w - 6)]$
3. $3(s^2 - 9)$
   $3[(s - 3)(s + 3)]$
4. $(u^5 - 10)(u^5 + 10)$
5. $49 - x^2$
   $(7 - x)(7 + x)$
6. $(z^6 + 9)(z^6 - 9)$
   $(z^6 + 9)(z^3 + 3)(z^3 - 3)$

Technique 4: Factoring Trinomials

In the previous section, when we multiplied two binomials such as $(x + 3)(x + 2)$, we got $x^2 + 2x + 3x + 6$, which is equal to $x^2 + 5x + 6$. Notice that the x^2 is the product of the first terms of the binomials. The $5x$ is the sum of the products of the inside and the outside terms of the binomial, and the 6 is the product of the last terms of the binomials. Now we'll undo this multiplication by factoring. When we factor a trinomial, as usual we need to first look for a G.C.F. Whenever we factor a trinomial it will be the product of two binomials. Now let's consider factoring the trinomial $x^2 + 5x + 6$. Factoring a trinomial is a kind of guessing game, but we can follow a few logical steps. First, since x^2 is the first term, we know that it's the product of x and x. So we can set up our factoring *this* way: $x^2 + 5x + 6 = (x\quad)(x\quad)$. Since the trinomial has no minus signs, there can't be any minus signs in either of the factors: $x^2 + 5x + 6 = (x+\quad)(x+\quad)$. All that's left is to find the numbers that we multiplied together that gave us 6, and that when added together give us 5. It doesn't take too many guesses to figure out that the numbers are 3 and 2.

$x^2 + 5x + 6 = (x + 3)(x + 2)$

Example 48:

Factor $x^2 - 2x - 24$.

Solution:

$(x\quad)(x\quad)$	We know we need x times x to get x^2.
$(x+\quad)(x-\quad)$	We need a + and a − to get a product of −24.
$(x - 6)(x + 4)$	We're looking for two numbers that will multiply to be −24 and combine to be −2. Those numbers are −6 and 4.

Example 49:

Factor $3c^2 + 7cd - 20d^2$.

$(3c\quad)(c\quad)$ We know we need $3c$ times c to get $3c^2$. When the leading coefficient is not a 1, we'll put the signs in last. We need two numbers that will multiply to be −20. We also know that the product of the inside and the outside terms must combine to be 7. We have to include d in the last terms to get d^2. The factors of 20 are 1 and 20, 2, and 10, and 4 and 5. By trial and error we find $3c^2 + 7cd - 20d^2 = (3c - 5d)(c + 4d)$. To check this we'll observe that the product of the last terms, $(-5d)(4d) = -20d^2$, and the sum of the products of the inside terms $(-5d)(c) = -5cd$ and the outside terms $(3c)(4d) = 12cd$ combine to equal $-5cd + 12cd = 7cd$. You can always check your factoring by multiplying the polynomials to see if you get the original trinomial.

SELF-TEST 11:

Factor the following:

1. $c^2 + c - 12$
2. $3d^2 - 18d + 24$
3. $2h^2 - 7h - 15$
4. $12k^2 + 2k - 24$
5. $30 - a - a^2$
6. $5x^2 + 10xy + 5y^2$
7. $3s^2 + 7st + 2t^2$
8. $3x^2 + 10xy + 8y^2$

ANSWERS:

1. $(c - 3)(c + 4)$
2. $3(d^2 - 6d + 8)$
   $3(d - 4)(d - 2)$
3. $(2h + 3)(h - 5)$
4. $2(6k^2 + k - 12)$
   $2(3k - 4)(2k + 3)$
5. $(5 - a)(6 + a)$
6. $5(x^2 + 2xy + y^2)$
   $5(x + y)(x + y)$
   $5(x + y)^2$
7. $(s + 2t)(3s + t)$
8. $(x + 2y)(3x + 4y)$

Technique 5: Sum/Difference of Two Cubes

$(x^3 - y^3) = (x - y)(x^2 + xy + y^2)$
$(x^3 + y^3) = (x + y)(x^2 - xy + y^2)$

The factored form of the sum or difference of two cubes is a binomial times a trinomial. The binomial has the same sign as the sum or difference of the cubes. The terms of the binomial are the cube root of the original terms. The first term of the trinomial is the square of the first term of the binomial, the second term of the trinomial is the product of the terms of the binomial, and the last term of the trinomial is the square of the second term of the binomial. The first term of the trinomial is always positive, the sign of the second term is always the opposite of the second term of the binomial, and the sign of the last term is always positive. If this sounds a little confusing, don't worry. As soon as we work out a few examples together, it'll be easy.

Example 50:

Factor $y^3 - 8$.
$(y - 2)(y^2 + 2y + 4)$

The cube root of y^3 is y, the cube root of 8 is 2. The binomial is $(y - 2)$. To get the trinomial, we square the first term of the binomial y and get y^2. We get the second term of the trinomial by multiplying the two terms of the binomial $(y)(2)$ and get $2y$. If the binomial is a difference, we make the second term, $2y$, a positive. To find the third term of the trinomial, we square the second term of the binomial: $(2)^2 = 4$.

Example 51:

Factor $x^6 + 27$.

Solution:

$(x^2 + 3)(x^4 - 3x^2 + 9)$ The cube root of x^6 is x^2; the cube root of 27 is 3; x^2 squared is x^4; 3 squared is 9; 3 times x^2 is $3x^2$. The binomial uses a + because the original example has a +. The trinomial is always the opposite sign.

SELF-TEST 12:

Factor the following:

1. $a^3 - 64$
2. $3a^3 - 81$
3. $10s^3 + 270$
4. $x^6 + 1$
5. $64y^9 - 125z^{12}$
6. $u^3 + v^3w^6$

ANSWERS:

1. $(a - 4)(a^2 + 4a + 16)$
2. $3(a^3 - 27) = 3(a - 3)(a^2 + 3a + 9)$
3. $10(s^3 + 27) = 10(s + 3)(s^2 - 3s + 9)$
4. $(x^2 + 1)(x^4 - x^2 + 1)$
5. $(4y^3 - 5z^4)(16y^6 + 20y^3z^4 + 25z^8)$
6. $(u + vw^2)(u^2 - uvw^2 + v^2w^4)$

5   Basic Geometry

In this section we will review some of the basic geometric formulas you will need to use in precalculus. Let's start with the distance formula and the midpoint formula for the line that passes through two points, (x_1, y_1), (x_2, y_2).

Distance formula:
$$d = \sqrt{(x_2 - x_1)^2 + (y_2 - y_1)^2}$$

Midpoint formula:
$$\left(\frac{x_1 + x_2}{2}, \frac{y_1 + y_2}{2}\right)$$

Example 52:

Find the distance between the points (2,3) and (5,7) and the midpoint of the line segment.

Solution:

$$d = \sqrt{(2-5)^2 + (3-7)^2} = \sqrt{(-3)^2 + (-4)^2} = \sqrt{9+16} = \sqrt{25} = 5$$

Midpoint:

$$\left(\frac{2+5}{2}, \frac{3+7}{2}\right) = \left(\frac{7}{2}, \frac{10}{2}\right) = \left(\frac{7}{2}, 5\right)$$

Now it's time to review some perimeter and area formulas for basic shapes.

Shape	Perimeter	Area
Square 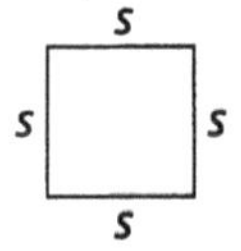 (sides s)	$P = 4s$	$A = s^2$
Rectangle 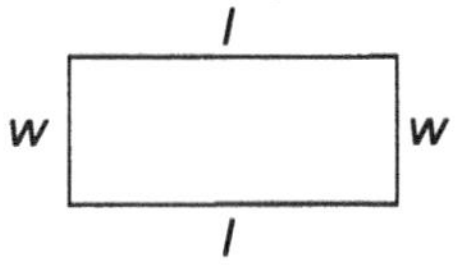 (length l, width w)	$P = 2l + 2w$	$A = lw$
Triangle 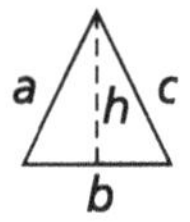 (sides a, b, c, height h)	$P = a + b + c$	$A = \frac{1}{2}bh$
Parallelogram 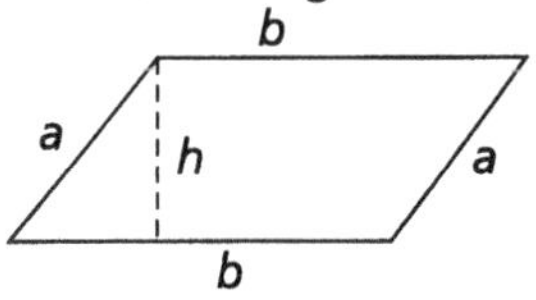 (sides a, b, height h)	$P = 2b + 2a$	$A = bh$

The perimeter of a circle is called the circumference (C) of the circle.

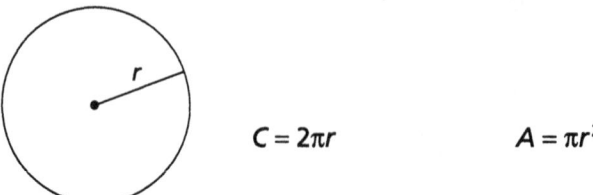

$C = 2\pi r$ $\qquad A = \pi r^2$

Let's not forget about the Pythagorean theorem, which applies to right triangles.

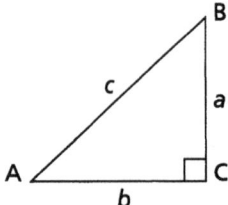

The theorem states that the square of the hypotenuse is equal to the sum of the square of the sides, or $c^2 = a^2 + b^2$.

Now that we have all the formulas we need, let's apply them to some examples.

Example 53:

Find the perimeter and the area of a square with sides 4 inches in length.

Solution:

$P = 4s = 4(4) = 16$ inches $\qquad A = s^2 = 4^2 = 16$ square inches

Example 54:

Find the perimeter and the area of a rectangle with length 5 feet and width 3 feet.

Solution:

$P = 2l + 2w = 2(5) + 2(3) = 16$ feet $\quad A = lw = 5(3) = 15$ square feet

Example 55:
Find the perimeter and the area of the triangle below:

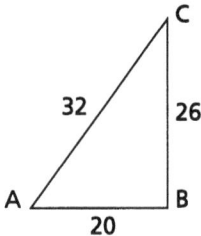

Solution:

$P = a + b + c = 26 + 32 + 20 = 78$ $A = \frac{1}{2} bh = \frac{1}{2}(20)(26) = 260$

Example 56:
Find the perimeter and the area of the parallelogram below.

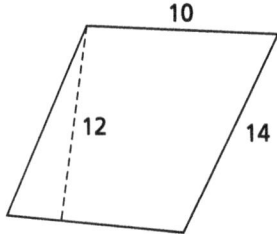

$P = 2b + 2a = 2(10) + 2(12) = 44$ $A = bh = 10(12) = 120$

Example 57:
Find the circumference and the area of a circle with a radius of 6 inches.

$C = 2\pi r = 2\pi(6) = 12\pi$ inches $A = \pi r^2 = \pi(6)^2 = 36\pi$ square inches

SELF-TEST 13:

Find the perimeter and the area of:

1. a square with side 10 inches
2. a rectangle with length 5 inches and width 7 inches
3. a parallelogram with side 3 inches, base 6 inches, and height 2 inches
4. the triangle with equal sides of length 4

28 PRECALCULUS

5. a circle of radius 12 inches

6. Find the distance of the line segment that connects the points $(-1, -2)$ and $(-4, -5)$.

ANSWERS:

1. $P = 4s = 4(10) = 40$ inches
$A = s^2 = 10^2 = 100$ square inches

2. $P = 2l + 2w = 2(5) + 2(7) = 24$ inches
$A = lw = 5(7) = 35$ square inches

3. $P = 2b + 2a = 2(6) + 2(3) = 18$ inches
$A = bh = 6(2) = 12$ square inches

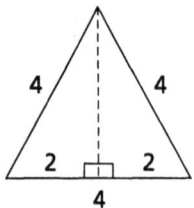

4. $P = a + b + c = 12$ To find the area, we need to find the height of the triangle.

We'll use the Pythagorean theorem to find h, and we'll use half of the triangle because the Pythagorean theorem applies to right triangles only.

$4^2 = h^2 + 2^2$, $4^2 - 2^2 = h^2$, $h^2 = 12$, $h = 2\sqrt{3}$

$A = \frac{1}{2}bh = \frac{1}{2}(4)(2\sqrt{3}) = 4\sqrt{3}$

5. $C = 2\pi r = 2\pi(12) = 24\pi$ inches
$A = \pi r^2 = \pi(12)^2 = 144\pi$ square inches

6. $d = \sqrt{(-1 - -4)^2 + (-2 - -5)^2}$
$d = \sqrt{3^2 + 3^2} = \sqrt{18} = 3\sqrt{2}$

Midpoint: $\left(\frac{-1-4}{2}, \frac{-2-5}{2}\right) = \left(\frac{-5}{2}, \frac{-7}{2}\right)$

6 Applications

In this section we will work on everybody's favorite topic, word problems. The hardest part about solving word problems is translating words into mathematical symbols. We'll work out the typical types you really need to know for precalculus. If you're one of those people who have a fear of word problems, don't worry—we'll make it as pain-free as possible. Let's get started.

Example 58:

Find the dimensions of a rectangle whose length is twice its width if its area is 50 square feet.

Solution:

We know from the previous section that the formula for area is $A = l(w)$. We're told that the length is twice the width. Stated in algebraic terms, we would say $l = 2w$.

$A = l(w)$
$50 = 2w(w)$
$50 = 2w^2$
$25 = w^2$
$w = 5$
$l = 2w = 2(5) = 10$

We'll substitute 50 for the A and $2w$ for the l. Next, we'll solve for the width w by dividing both sides of the equation by 2 and then finding the $\sqrt{25}$. Once we know the value of w, we multiply it by 2 to find the length.

The dimensions of the rectangle are 10 by 5.

Example 59:

A collection of 56 coins consisting of only nickels and dimes has a value of $4.00. How many nickels are there?

Solution:

Our unknown is the number of nickels, so we'll let x equal the number of nickels. Let $56 - x$ equal the number of dimes, because there are a total of 56 coins, and if we take out all the nickels, we should be left with only the dimes. Four dollars is the value of the coins, not the number of coins. To represent the value of the nickels we'll use $.05x$; the value of the dimes will be $.10(56 - x)$.

The total value of the coins equals the value of the dimes plus the value of the nickels.

$\$4.00 = .10(56 - x) + .05x$ Move all the decimals two places to the right.
$400 = 10(56 - x) + 5x$ Distribute the 10 and subtract 560 from each side and
$400 = 560 - 10x + 5x$ combine the $-10x$ and the $5x$.
$-160 = -5x$ Divide both sides by -5.
$32 = x$

There are 32 nickels.

Example 60:

A total of $5,000 is invested in two accounts. One investment earned 4.6% annual simple interest, while the other investment earned 5.5%.

Total earnings from both investments were $254.75. Find the amount invested at 4.6%.

Solution:

We'll let x equal the amount invested at 4.6%. The amount invested at 5.5% would be the total amount of money minus the amount invested at 4.6%, so $5,000 - x$ equals the amount invested at 5.5%. Total amount of interest equals the amount of interest earned at 4.6% plus the interest earned at 5.5%.

$254.75 = .046x + .055(5000 - x)$	Move the decimals three places to the right.
$254750 = 46x + 55(5000 - x)$	Distribute the 55.
$254750 = 46x + 275000 - 55x$	Subtract 275000 from both sides.
$-20250 = -9x$	Combine the $-55x$ and the $46x$.
$2250 = x$	Divide both sides by -9.

A total of $2,250 was invested at 4.6%.

SELF-TEST 14:

1. Find the side s of a square with diagonal 10 inches.
2. The perimeter of a rectangle is 24 feet. If the length is three more than twice the width, find the length.
3. The sum of three consecutive odd integers is 55 more than the sum of the first two. Find the numbers.
4. A 20-foot ladder is leaning against a building. The bottom of the ladder is 12 feet from the base of the building. How high up the building does the ladder reach?
5. What number must be added to the numerator and the denominator of $\frac{5}{7}$ to produce the fraction $\frac{4}{5}$?
6. Two cars 225 miles apart are traveling toward each other. One car travels twice as fast as the other. The cars meet in 2.5 hours. How fast are the cars traveling?

ANSWERS:

1. We can divide the square into two right triangles, then use the Pythagorean theorem.

The Basics 31

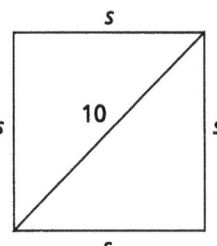 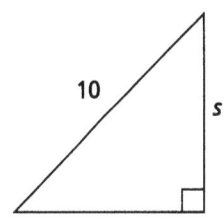

$a^2 + b^2 = c^2$
$s^2 + s^2 = 10^2$
$2s^2 = 100$
$s^2 = 50$
$s = \sqrt{50} = 5\sqrt{2}$

This is an isosceles triangle.

2. $P = 2l + 2w$, $P = 24$, $l = 3 + 2w$
$24 = 2(3 + 2w) + 2(w)$
$24 = 6 + 4w + 2w$
$18 = 6w$
$w = 3$, $l = 3 + 2w = 3 + 2(3) = 9$

3. $x + x + 2 + x + 4 = 55 + x + x + 2$
$3x + 6 = 57 + 2x$

$x = 51$

We'll call the first odd integer x.
To get to the next odd integer we'd have to move two units to the right on the number line, so we'll say the next odd integers are $x + 2$ and $x + 4$.

The consecutive odd integers are 51, 53, and 55.

4. We'll use the Pythagorean theorem because this is a right-triangle problem.
$20^2 = 12^2 + x^2$
$400 = 144 + x^2$
$256 = x^2$
$16 = x$

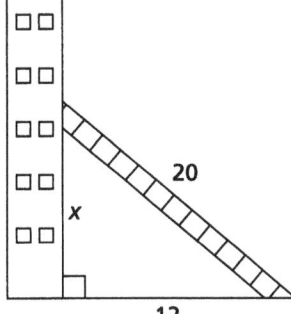

5. We have to add a number to the numerator and the denominator of $\frac{5}{7}$.

$\frac{5 + x}{7 + x} = \frac{4}{5}$
$5(5 + x) = 4(7 + x)$
$25 + 5x = 28 + 4x$
$x = 3$

6. This is a distance problem, so we'll have to use our formula distance = rate times time $D = R(T)$.

The cars are traveling toward each other. The total distance covered by both cars is 225 miles. Let's let R represent the rate of the slower car and $2R$ the rate of the faster car. The time they're traveling is 2.5 hours. The distance the slower car travels is $2.5R$. The distance the faster car travels is $2.5(2R)$.

Total distance = distance traveled by the slower car + distance traveled by the faster car.
$225 = 2.5R + 2.5(2R)$
$2250 = 25R + 50R$
$2250 = 75R$

$R = 30$ miles per hour; the rate of the faster car is $2R = 2(30) = 60$ miles per hour.

2 Functions

1 Definition of a Function

Functions are arguably the most important mathematical equations and are central to calculus. When you've completed this chapter you should be able to:

- define a function and state its domain and range
- perform operations on functions
- form composite functions
- find the inverse of a function

Functions have many real-life applications. Whenever we find that two things are related to each other by some type of rule or correspondence, it's a function. For example, the amount of weight you gain is a function of how much food you eat. The distance you drive is a function of how fast you drive. The amount of money you make is a function of how many hours you work.

So just what *is* a function? Before we give you a formal mathematical definition of a function, we'll explain it in simple English: A function is nothing more than a set of directions, where one value is dependent on another. For example, if you were driving a car and we instructed you to push your foot down harder on the accelerator pedal, the speed of the car would increase. The speed of the car is dependent on the pounds of pressure you put on the accelerator pedal. The amount of pressure you put on the accelerator is called the independent variable, because you decide how much pressure to use, and the speed is called the dependent variable, because the speed at which you travel depends on the amount of pressure you choose to put on the accelerator. One other important thing you need to know about the definition of a function is that for every independent variable there is one and only one corresponding dependent variable. For every amount of pressure you put on the accelerator the car will go a specific speed. For example, if you put two pounds of pressure on your accelerator your car might go 40 miles an hour; it will not go 80 miles an hour. For each independent variable there is one and only one corresponding dependent variable.

Now it's time for a formal mathematical definition of a function.

A function is a correspondence between a first set of independent variables, called the domain, and a second set of dependent variables, called the range, such that for each member of the domain there corresponds exactly one member of the range. In other words, for each x value there corresponds one y value.

If an independent variable has more than one corresponding variable, it's called a relation. For example, if we listed the clients of a bank with the balance of their bank accounts, some people would have more than one account and therefore more than one balance.

A relation is a correspondence between a first set and a second set, such that for some members of the first set there corresponds at least one member of the second set.

One way to show this correspondence is to use what's called a mapping from the domain to the range. Another way is to use function notation. The next few examples will illustrate both of these techniques, and while we're at it, we will also ask you if each example is a function or a relation.

Example 1:

$y = x^2$

This is a function, because for every value for x we get one and only one value for y. This function tells us to take whatever independent variable we give it and raise it to the second power. To present $y = x^2$ in function notation we would write $f(x) = x^2$, or $g(x) = x^2$, or $h(x) = x^2$, etc. The left side of the first equation is read "f of x." It means y is a function of x.

The x is the independent variable and the y is the dependent variable. The domain is the set of all the independent values, x. The range is the set of all the dependent values, y. If we follow the set of directions for this function, all we do is raise whatever we substitute inside the function to the second power. If we substitute the number 0, we would raise 0 to the second power and get 0. We would show this by using function notation in the following way: $f(0) = 0^2 = 0$. The same is true for -1, 1, 2, -2, etc. $f(1) = 1^2 = 1$, $f(-1) = (-1)^2 = 1$, $f(2) = 4$, $f(-2) = 4$, etc. The left figure below is a mapping of the function $f(x) = x^2$. The right side shows the graph of $f(x) = x^2$. Notice that for every x value on the graph there is one and only one corresponding y value. If you can draw a vertical line through any part of a graph and if it intersects the graph at only one point, it's a function. This shows us that for each x value there is one and only one y value. This is called the vertical-line test.

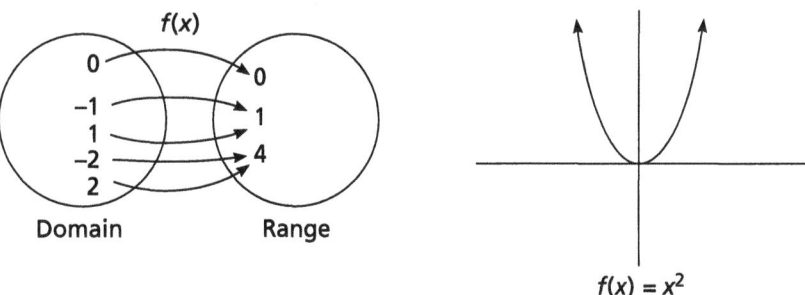

Example 2:

$y = \pm\sqrt{x}$ Let's look at the mapping and the graph of $y = \pm\sqrt{x}$ to decide if $y = \pm\sqrt{x}$ is a function or a relation.

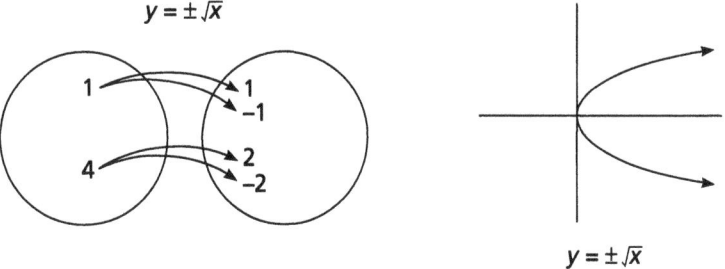

If you said it's a relation, not a function, because there are two y values for one x value, you're learning. Or you could have said it's not a function because its graph fails the vertical-line test.

Example 3:

$f(x) = -2x^2 + 3x - 1$ Find

a. $f(2)$ **b.** $f(-3)$ **c.** $f(a+b)$

Solution:

This set of directions tells us to take whatever we give it and raise it to the second power, multiply that result by –2, then add three times the original value, then subtract 1.

a. $f(2) = -2(2)^2 + 3(2) - 1 = -3$
b. $f(-3) = -2(-3)^2 + 3(-3) - 1 = -28$
c. $f(a+b) = -2(a+b)^2 + 3(a+b) - 1$
    $-2(a+b)(a+b) + 3(a+b) - 1$
    $-2(a^2 + 2ab + b^2) + 3a + 3b - 1$
    $-2a^2 - 4ab - 2b^2 + 3a + 3b - 1$

This is the first problem where we're substituting an algebraic expression instead of a number. We still follow our set of directions in the same way.

Now it's time for you to try a few problems.

SELF-TEST 1:

In problems 1 to 4, decide whether the given is a function or a relation. In problems 5 to 8, find

a. $f(1)$ **b.** $f(-1)$ **c.** $f(x^3)$ **d.** $f(x+y)$

1.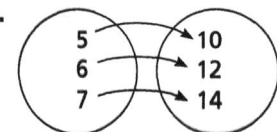

2.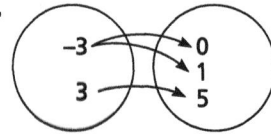

3.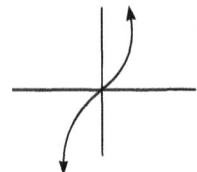

4.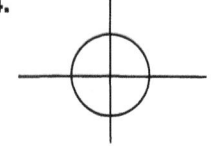

5. $f(x) = 5x + 9$

6. $f(x) = -2x^2 + 5x - 3$

7. $f(x) = 4\sqrt{x}$

8. $f(x) = \dfrac{x^2 - 2}{x + 3}$

Functions 37

ANSWERS:

1. This is a function; for every x value there is one corresponding y value.
2. This is a relation, not a function, because the x value –3 has two corresponding y values.
3. This is a function; it passes the vertical-line test. Any vertical line drawn through this graph intersects at only one point.
4. This is a relation, not a function, because it fails the vertical-line test.
5. a. $f(1) = 5(1) + 9 = 14$
   b. $f(-1) = 5(-1) + 9 = 4$
   c. $f(x^3) = 5(x^3) + 9 = 5x^3 + 9$
   d. $f(x+y) = 5(x+y) + 9 = 5x + 5y + 9$
6. a. $f(1) = -2(1)^2 + 5(1) - 3 = 0$
   b. $f(-1) = -2(-1)^2 + 5(-1) - 3 = -10$
   c. $f(x^3) = -2(x^3)^2 + 5(x^3) - 3 = -2x^6 + 5x^3 - 3$
   d. $f(x+y) = -2(x+y)^2 + 5(x+y) - 3$
      $-2(x^2 + 2xy + y^2) + 5x + 5y - 3$
      $-2x^2 - 4xy - 2y^2 + 5x + 5y - 3$
7. a. $f(1) = 4\sqrt{1} = 4$
   b. $f(-1) = 4\sqrt{-1}$ is not a real number
   c. $f(x^3) = 4\sqrt{x^3} = 4x\sqrt{x}$
   d. $f(x+y) = 4\sqrt{x+y}$
8. a. $f(1) = \dfrac{1^2 - 2}{1 + 3} = -\dfrac{1}{4}$
   b. $f(-1) = \dfrac{(-1)^2 - 2}{-1 + 3} = -\dfrac{1}{2}$
   c. $f(x^3) = \dfrac{(x^3)^2 - 2}{x^3 + 3} = \dfrac{x^6 - 2}{x^3 + 3}$
   d. $f(x) = \dfrac{(x+y)^2 - 2}{x+y+3} = \dfrac{x^2 + 2xy + y^2 - 2}{x+y+3}$

Now that you have a better understanding of the difference between a function and a relation, let's take a closer look at acceptable members of the domain and range of a function. Finding the restrictions on the domain of a function is a very important topic. Let's start by going back to example 1, $f(x) = x^2$. The domain is the set of values we substitute in the function for x. The first question we should ask ourselves is, "Is there any number we cannot square?" The answer is no; we can raise any number to the second power. So our domain is the set of all real numbers. We'll use the symbol $\Re$ to represent the set of all real numbers. There are no restrictions on the domain of this function.

We also need to talk about the range, which is the set of all the numbers we get back when we substitute the values of the domain into the function. What kind of numbers would we get back when we square a number? We wouldn't get any negative numbers, only positives and 0. Our range would be the set of all nonnegative numbers. We would write that in interval notation as $[0,\infty)$. Just a reminder: When using interval notation, a [or] means the number is included in the set; a (or) means the number is excluded. For example (3,6] means the set of numbers between 3 and 6, excluding the 3 and including the 6. If there is no last number in our set we use the symbols $-\infty$ or ∞.

Example 4:

State the domain and range of the function $g(x) = \dfrac{x^3}{x^2 - 1}$.

Solution:

Let's ask ourselves if there is any number that x can't be. At first it appears as though we could substitute any value for x, but, on closer inspection, we really can't. Suppose we substituted a 1 for x. Let's see what happens.

$$f(1) = \dfrac{(1)^3}{(1)^2 - 1} = \dfrac{1}{0}$$

As we know, division by 0 is undefined, so 1 can't be in our domain. The same is true of −1. You may ask, "How did you figure out 1 and −1 were the magic numbers?" Whenever we have a fraction, we know the denominator can't equal 0, so we set the denominator equal to 0 and solve for x.

$x^2 - 1 = 0$	Factor as the difference of two squares.
$(x - 1)(x + 1) = 0$	
$x - 1 = 0, x + 1 = 0$	Set each factor to 0 and solve for x.
$x = 1, x = -1$	Our domain is the set of all numbers except 1 and −1. We'll write that using interval notation. Domain: $x \neq -1, x \neq 1$. $(-\infty, -1) \cup (-1, 1) \cup (1, \infty)$

Let's find the range. Is there any number we can't get back when we substitute a value for x? No; we can get negatives, 0, and positives, so our range is the set of all real numbers. Range: $\Re$.

Example 5:

Find the domain and range of $h(x) = \sqrt{x - 2}$.

Solution:

To find the domain we should keep in mind that we can't take an even root of a negative. Whatever's under the radical has to be at least a 0.

Let's use an inequality to figure out what values of x will make $x - 2$ at least a 0.

$x - 2 \geq 0$
$x \geq 2$ This gives us our domain of $[2, \infty)$.

Let's find the range. When we take the square root of a number, we don't get back a negative; we can get back a 0 or a positive number. Range: $[0, \infty)$.

SELF-TEST 2:

Find the domain and range of the following functions:

1. $f(x) = 2x + 1$
2. $g(x) = x^3$
3. $h(x) = \dfrac{x}{x}$
4. $t(x) = \sqrt{x^2 - 4}$
5. $s(x) = |x|$
6. $f(x) = \sqrt{\dfrac{x - 2}{x - 1}}$

ANSWERS:

When we write our solutions we'll also show the graph of the function. This will help us to determine the domain and range of the function. Remember: The domain is the set of all the possible x values of the function, and the range is the set of all the y values of the function. The x values measure the horizontal distance (left to right) from the origin. The origin is the point where the x and the y axes intersect. Its coordinates are (0,0). If we want to determine the members of the domain we look to see if the x values have any restrictions on them. The y values measure the vertical distance (up and down) from the origin. If we want to determine the members of the range, we look to see if the y values have any restrictions on them.

1.
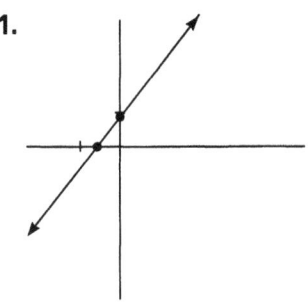

$fx = 2x + 1$

We can multiply any number by 2, then add 1. If we look at the graph of the function we see that the x values can go all the way to the left or the right on the x-axis. Domain: $\Re$. When we substitute numbers for x we can get back any number for y. If we look at the graph of the function we see that the y values can go all the way up or down on the y-axis. Range: $\Re$.

2.

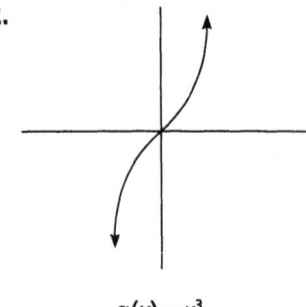

$g(x) = x^3$

We can put in any value for x. There is no restriction on how far to the right or the left the graph can go. Domain: ℜ.

We can get back any value for y. There is no restriction on how far up or down the graph can go. Range: ℜ.

3.

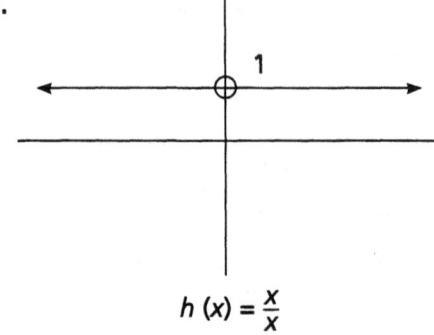

$h(x) = \dfrac{x}{x}$

We can substitute any value for x. There is no restriction on how far to the right or the left the graph can go. Domain: ℜ. $x \neq 0$.

Because anything divided by itself is always 1, the only y value we could ever get is 1. The only exception to this rule is: 0 divided by 0 is an indeterminate form. You will not study indeterminate forms until calculus II. The graph shows us that this is the horizontal line $y = 1$. Range: $y = 1$.

4. $t(x) = \sqrt{x^2 - 4}$

Finding the domain of this function is a little more work than for the ones we've done so far. We can't take the even root of a negative, so we know $x^2 - 4$ has to be at least zero.

$x^2 - 4 \geq 0$
$(x - 2)(x + 2)$
$x = 2, x = -2$
$(-\infty, -2]\ [-2, 2]\ [2, \infty)$

From intermediate algebra you know that we first solve to find where y or $t(x) = 0$. Then we partition the number line into intervals and substitute values from each interval into $\sqrt{x^2 - 4}$ to find our domain: $(-\infty, -2] \cup [2, \infty)$. The $[-2, 2]$ is not part of the domain because if we substitute a value from that interval into the function the result would be the square root of a negative number. The range would have to be all nonnegative numbers. Range: $[0, \infty)$.

5.

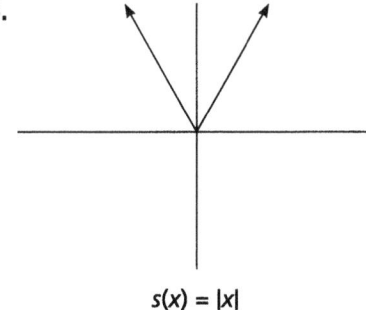

$s(x) = |x|$

The domain of this function is all real numbers because we can take the absolute value of any number. Notice that the graph has no limit on how far to the right or the left it can go. When we take the absolute value of a number we never get a negative number. Range: [0,∞). The graph doesn't have a limit on how high it can go, but it never goes below 0.

6.

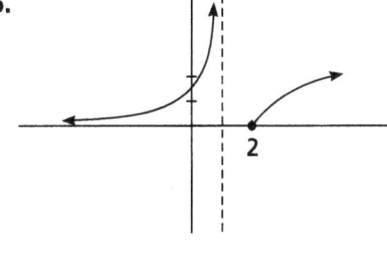

$f(x) = \sqrt{\dfrac{x-2}{x-1}}$

This is our hardest problem. If you can do this, you can do anything. We know we can only put in values for x that will make the fraction under the radical nonnegative, so we know $\dfrac{x-2}{x-1}$ must be at least 0. We also know that the only time a fraction is at least 0 is when the numerator and the denominator are both negative or both positive. We also know that the denominator x − 1 can't equal 0 because that would result in an undefined fraction. This fraction would equal 0 when the numerator x − 2 equals 0, which occurs when x = 2, so we'll divide the number line around 1 and 2. This would create the following intervals: (−∞,1), (1,2], and [2,∞). If we substitute values into f(x) from each interval to see which ones are acceptable, we find our domain of (−∞,1) ∪ [2,∞). The lowest value we could get for y is 0, and there is no limit on how high the y values can go. Range: [0,∞).

2 Operations on Functions

In arithmetic we learn the basic operations of addition, subtraction, multiplication, and division. These basics also apply to functions. In this section we will learn how to add, subtract, multiply, and divide functions. Let's get started!

Example 6:
Given the functions $f(x) = 2x + 1$, $g(x) = 4x^2$, and $h(x) = \sqrt{x}$, find

a. $f(x) + g(x)$ **b.** $-h(x)$ **c.** $h(-x)$ **d.** $\dfrac{f(x)}{g(x)}$ **e.** $g(x) \cdot h(x)$

Solutions:

a. $2x + 1 + 4x^2 = 4x^2 + 2x + 1$
b. The negative is outside the function, not inside; $-h(x) = -\sqrt{x}$.
c. To find $h(-x)$ we substitute $-x$ into the function and get $\sqrt{-x}$.
d. $\dfrac{2x + 1}{4x^2}$

e. $4x^2(\sqrt{x}) = 4x^2\sqrt{x}$

Let's try a few more, but this time let's also substitute some numbers inside the functions.

Example 7:
Given the functions $f(x) = x^3$, $g(x) = x^2 + 2x + 1$, $j(x) = x + 1$, find

a. $f(2) + g(3)$ **b.** $\dfrac{g(x)}{j(x)}$ **c.** $j(1) \cdot f(-2)$

d. $-j(2) - g(0)$ **e.** $[j(3)]^2$

Solutions:

a. $f(2) = 2^3 = 8 \quad g(3) = (3)^2 + 2(3) + 1 = 16 \quad f(2) + g(3) = 8 + 16 = 24$

b. $\dfrac{g(x)}{j(x)} = \dfrac{x^2 + 2x + 1}{x + 1} = \dfrac{(x + 1)(x + 1)}{x + 1} = x + 1$

c. $j(1) = 1 + 1 = 2, \quad f(-2) = (-2)^3 = -8, \quad j(1) \cdot f(-2) = 2(-8) = -16$
d. $-j(2) = -(2 + 1) = -3, \quad g(0) = 0^2 + 2(0) + 1 = 1, \quad -j(2) - g(0) = -3 - 1 = -4$
e. $j(3) = 3 + 1 = 4, \quad [j(3)]^2 = [4]^2 = 16$

SELF-TEST 3: Given $f(x) = x + 2$, $g(x) = 2x^2$, $h(x) = \dfrac{x + 1}{x - 1}$, $i(x) = \dfrac{1}{3}x$, and $j(x) = 2x + 4$, $k(x) = \sqrt{x}$, find

1. $f(2) - g(3)$ **2.** $g(x^2)$ **3.** $[g(x)]^2$ **4.** $\dfrac{f(x)}{g(x)}$

5. $k(16) + 5i(3)$ **6.** $\dfrac{f(x + 1)}{3j(x)}$ **7.** $5k(x^2) - g(4x + 1)$

8. $\dfrac{1}{h(x)}$ 9. $\left(\dfrac{\frac{1}{f(x)}}{j(x)}\right)^2$ 10. $k(25) - i(-9)$

ANSWERS:

1. $f(2) = 2 + 2 = 4$ $g(3) = 2(3)^2 = 18$ $f(2) - g(3) = 4 - 18 = -14$

2. $g(x^2) = 2(x^2)^2 = 2x^4$ 3. $[g(x)]^2 = [2x^2]^2 = 4x^4$ 4. $\dfrac{f(x)}{g(x)} = \dfrac{x+2}{2x^2}$

5. $k(16) = \sqrt{16} = 4$ $5i(3) = 5\left[\dfrac{1}{3}(3)\right] = 5$ $k(16) + 5i(3) = 4 + 5 = 9$

6. $f(x + 1) = x + 1 + 2 = x + 3$ $3j(x) = 3(2x + 4) = 6x + 12$

    $\dfrac{f(x+1)}{3j(x)} = \dfrac{x+3}{6x+12}$

    $\dfrac{f(x+1)}{3j(x)} = \dfrac{x+3}{6x+12}$

7. $5k(x^2) = 5\sqrt{x^2} = 5x$ $g(4x + 1) = 2(4x + 1)^2 = 2(16x^2 + 8x + 1) = 32x^2 + 16x + 2$
   $5k(x^2) - g(4x + 1) = 5x - 32x^2 - 16x - 2 = -32x^2 - 11x - 2$

8. $\dfrac{1}{h(x)} = \dfrac{1}{\frac{x+1}{x-1}} = \dfrac{x-1}{x+1}$

9. $\dfrac{1}{\frac{f(x)}{j(x)}} = \dfrac{1}{\frac{x+2}{2x+4}} = \dfrac{1}{\frac{x+2}{2(x+2)}} = \dfrac{1}{\frac{1}{2}} = 2$ $\left(\dfrac{\frac{1}{f(x)}}{j(x)}\right)^2 = (2)^2 = 4$

10. $k(25) = \sqrt{25} = 5$ $i(-9) = \dfrac{1}{3}(-9) = -3$

    $k(25) - i(-9) = 5 - (-3) = 8$

3 Composite Functions

In the previous section we worked on operations involving functions. In this section we will work on one more operation involving functions. This operation is called composition of functions, where we substitute a function into another function. Suppose we had two functions $f(x) = x^2$ and $g(x) = x + 2$, and we wanted to substitute $g(x)$ into $f(x)$. We would do that by substituting $x + 2$ into $f(x)$ and would get $f(x + 2) = (x + 2)^2$, which is equal to $x^2 + 4x + 4$. We would call this the composite function of f and g of x. We would write that in one of two possible ways: $(f \circ g)(x)$ or $f(g(x))$. This is read f of g of x. In this example g is the inner function and f is the outer function. We always substitute the inner function into the outer function. Let's try reversing the functions and see what happens. This time let's find the composite function of g of f of x, $g(f(x))$. We substitute the inner function $f(x)$ into the outer function $g(x)$: $g(f(x)) = g(x^2) = x^2 + 2$. Composition of functions is not commutative—that is, $f(g(\underline{x})) \neq g(f(x))$.

Example 8:

Given $h(x) = x - 3$, and $j(x) = \sqrt{x}$, find:

a. $h(j(x))$ **b.** $j(h(x))$

Solutions:

a. We always substitute the inner function into the outer function; $j(x)$ is the inner function; $h(j(x)) = h(\sqrt{x}) = \sqrt{x} - 3$.

b. The inner function is $h(x)$. $j(h(x)) = j(x - 3) = \sqrt{x - 3}$.

Example 9:

Given $t(x) = 2x$, $s(x) = x^4$, find

a. $t(s(4))$ **b.** $s(t(4))$

Solutions:

a. $s(4) = 4^4 = 256 \quad t(256) = 512$ b. $t(4) = 8 \quad s(8) = 4{,}096$

Example 10:

Given $t(x) = x + 3$, $s(x) = x^2 - 3x + 2$, find $t(s(2))$.

Solution:

$s(2) = 2^2 - 3(2) + 2 = 0, \quad t(0) = 0 + 3 = 3$

SELF-TEST 4: Find the following composite functions, given $f(x) = x^3$, $h(x) = x - 9$, $p(x) = \dfrac{\sqrt{x} - 4}{x + 4}$, $v(x) = 5x - 3$.

1. $v(f(x))$ **2.** $h(p(16))$ **3.** $p(h(16))$ **4.** $v(p(x))$

5. $h(v(1))$ **6.** $v(h(1))$ **7.** $h(f(x))$ **8.** $v(h(f(x)))$

ANSWERS:

1. $v(f(x)) = v(x^3) = 5(x^3) - 3 = 5x^3 - 3$

2. $p(16) = \dfrac{\sqrt{16} - 4}{16 + 4} = \dfrac{4 - 4}{20} = 0 \quad h(p(16)) = h(0) = 0 - 9 = -9$

3. $h(16) = 16 - 9 = 7 \quad p(h(16)) = p(7) = \dfrac{\sqrt{7} - 4}{7 + 4} = \dfrac{\sqrt{7} - 4}{11}$

4. $v(p(x)) = v\left(\dfrac{\sqrt{x} - 4}{x + 4}\right) = 5\left(\dfrac{\sqrt{x} - 4}{\sqrt{x} + 4}\right) - 3$

5. $v(1) = 5(1) - 3 = 2$ $h(v(1)) = h(2) = 2 - 9 = -7$

6. $h(1) = 1 - 9 = -8$ $v(h(1)) = v(-8) = 5(-8) - 3 = -43$

7. $h(f(x)) = h(x^3) = x^3 - 9$

8. $v(h(f(x))) = v(h(x^3)) = v(x^3 - 9) = 5(x^3 - 9) - 3 = 5x^3 - 45 - 3 = 5x^3 - 48$

4   Inverse Functions

Addition and subtraction are inverse operations—they "undo" each other: $4 + 2 = 6$, $6 - 4 = 2$. Some functions are inverses of each other. For example, $f(x) = x + 4$ and $g(x) = x - 4$ are inverses of each other; $f(3) = 7$ and $g(7) = 3$. The following diagram shows the mapping of these functions: $f(x)$ maps 3 into 7, and $g(x)$ maps 7 back into 3. It also shows the graphs of these two functions. Notice on the graphs that the point (3,7) is on $y = x + 4$, and the point (7,3) is on its inverse function, $y = x - 4$. When functions are inverses of each other the members of the domain of $f(x)$ become the members of the range of $f^{-1}(x)$, and the members of the range of $f(x)$ become the members of the domain of $f^{-1}(x)$. In other words, the x values of $f(x)$ are the y values of $f^{-1}(x)$, and the y values of $f(x)$ are the x values of $f^{-1}(x)$.

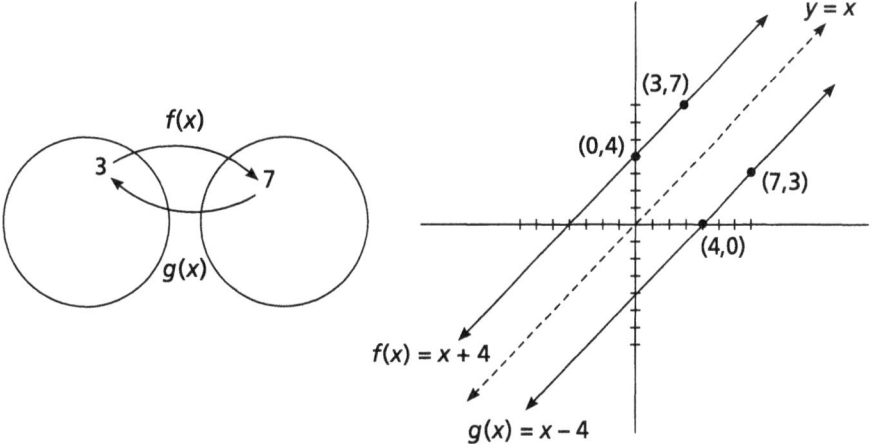

We've labeled some of the points on the graph so you can see how the x and the y coordinates switch position for $f(x)$ and $f^{-1}(x)$. The x and the y values swap. Notice that the graphs are reflections around the line $y = x$.

In this example we gave you the inverse function for $f(x)$. Suppose you had to find it for yourself. We would use the following easy step-by-step procedure to find an inverse function.

Steps to find an inverse function:

1. Replace $f(x)$ with y.
2. Interchange x and y.
3. Solve for y.
4. Replace y with $f^{-1}(x)$.

Mappings of inverse functions:

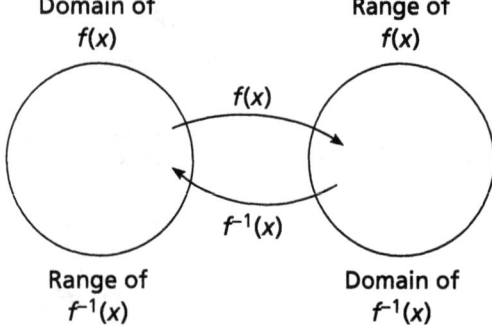

Example 11:

Find $f^{-1}(x)$ for $f(x) = 2x + 5$.

Solution:

$f(x)$ and y are interchangeable for all functions, so let's write this example as

$y = 2x + 5$	Replace $f(x)$ with y.
$x = 2y + 5$	Interchange x and y.
$x - 5 = 2y$	Solve for y.
$\dfrac{x-5}{2} = y$	
$f^{-1}(x) = \dfrac{x-5}{2}$	Replace the y with $f^{-1}(x)$.

In section 1 of this chapter we used the vertical-line test to see whether a statement was a relation or a function. If the vertical line intersects the graph in only one point it's a function, because for every x value a function should have only one y value. Now we're going to use the horizontal-line test to see whether a function has an inverse, which is also a function. If a horizontal line drawn through the graph intersects it in only one point, the function is called one to one, which is written 1–1. If a graph is not one to one, that means that two x values have the same y value and therefore the function's inverse is not a function.

Example 12:

Look at the following graph of $g(x) = x^2$. Is it a 1–1 function? Does it have an inverse function?

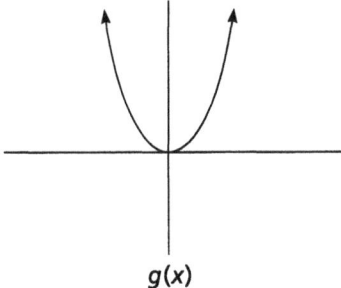

g(x)

Solution:

No, it's not a 1–1 function because it fails the horizontal-line test. Since two x values have the same y value, it does not have an inverse function.

Example 13:

Given $h(x) = x^3 + 2$, look at the following graph to determine whether $h(x)$ is 1–1 and therefore whether it has an inverse function. If it does, find $h^{-1}(x)$. Then look at the way $h(x)$ and $h^{-1}(x)$ reflect around the line $y = x$.

Solution:

Notice that the graph of $h(x)$ passes the vertical-line test.

$y = x^3 + 2$	Replace $h(x)$ with y.
$x = y^3 + 2$	Interchange the x and the y.
$x - 2 = y^3$	Solve for y.
$\sqrt[3]{x-2} = y$	
$h^{-1}(x) = \sqrt[3]{x-2}$	Replace y with $h^{-1}(x)$.

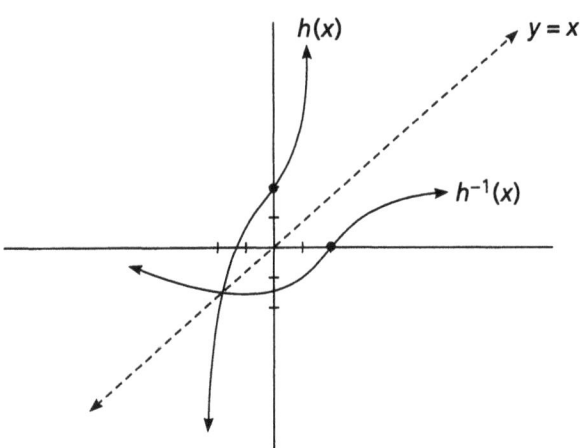

48   PRECALCULUS

If you go back to our earlier graphs of functions and their inverses, you would find that *the graph of f⁻¹ is a reflection of the graph of f across the line* y = x.

Example 14:
Given $f(x) = 5x + 8$, find $f^{-1}(x)$; find $f^{-1}(2)$.

Solution:

$y = 5x + 8$	Replace $f(x)$ with y.
$x = 5y + 8$	Interchange x and y.
$x - 8 = 5y$	Solve for y.
$\dfrac{x-8}{5} = y$	
$f^{-1}(x) = \dfrac{x-8}{5}$	Replace y with $f^{-1}(x)$.
$f^{-1}(2) = \dfrac{2-8}{5} = -\dfrac{6}{5}$	Substitute 2 into $f^{-1}(x)$.

SELF-TEST 5:

In problems 1 and 2, use the given graph of the function to state whether an inverse function exists. In problems 3 to 8, find the inverse function.

1.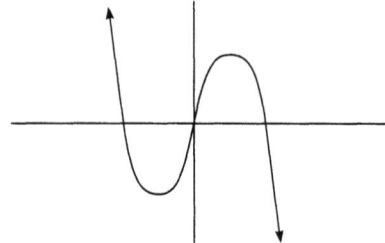

2.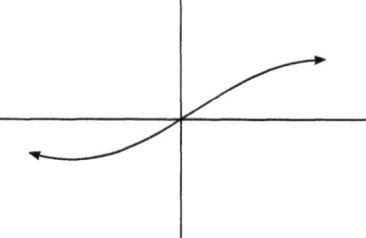

3. $f(x) = x^3 + 1$

4. $s(x) = 3x - 4$

5. $g(x) = \dfrac{2x - 3}{4}$

6. $h(x) = \sqrt[3]{x + 2} - 2$

7. $j(x) = -\dfrac{4}{x}$

   find $j^{-1}(4)$

8. $f(x) = \dfrac{x + 4}{x - 3}$

   find $f^{-1}(10)$

Functions

ANSWERS:

1. This is not a one-to-one function because a horizontal line intercepts the graph in more than one point. Two different x values have the same y value.

2. This is a one-to-one function because a horizontal line does not intercept the graph at more than one point.

3.
$y = x^3 + 1$
$x = y^3 + 1$
$x - 1 = y^3$
$x = \sqrt[3]{x-1}$
$f^{-1}(x) = \sqrt[3]{x-1}$

4.
$y = 3x - 4$
$x = 3y - 4$
$x + 4 = 3y$
$\dfrac{x+4}{3} = y$
$s^{-1}(x) = \dfrac{x+4}{3}$

5. $g(x) = \dfrac{2x-3}{4}$
$y = \dfrac{2x-3}{4}$
$x = \dfrac{2y-3}{4}$
$\dfrac{4x+3}{2} = y$
$g^{-1}(x) = \dfrac{4x+3}{2}$

6.
$y = \sqrt[3]{x+2} - 2$
$x = \sqrt[3]{y+2} - 2$
$x + 2 = \sqrt[3]{y+2}$
$(x+2)^3 - 2 = y$
$h^{-1}(x) = (x+2)^3 - 2$

7. $y = -\dfrac{4}{x}$
$x = -\dfrac{4}{y}$
$xy = -4$
$y = -\dfrac{4}{x}$
$j^{-1}(x) = -\dfrac{4}{x}$
$j^{-1}(4) = -1$

8. $y = \dfrac{x+4}{x-3}$
$x = \dfrac{y+4}{y-3}$
$xy - 3x = y + 4$
$xy - y = 3x + 4$
$y(x-1) = 3x + 4$
$y = \dfrac{3x+4}{x-1}$
$f^{-1}(x) = \dfrac{3x+4}{x-1}$
$f^{-1}(10) = \dfrac{34}{9}$

Suppose we gave you two functions and wanted you to verify that they're each other's inverse. There's a very simple way to do that. If we figure out their composite functions and we get an x, they're each other's inverse function.

If f(x) is a one-to-one function, then $f^{-1}(f(x)) = x = f(f^{-1}(x))$.

Example 15:

Are the functions $f(x) = -2x + 3$ and $g(x) = -\frac{1}{2}x + \frac{3}{2}$ inverses of each other?

Solution:

We'll use the property $f^{-1}(f(x)) = x$ and $f(f^{-1}(x)) = x$.

$$f(g(x)) = f\left(-\frac{1}{2}x + \frac{3}{2}\right) = -2\left(-\frac{1}{2}x + \frac{3}{2}\right) + 3 = x - 3 + 3 = x$$

$$g(f(x)) = g(-2x + 3) = -\frac{1}{2}(-2x + 3) + \frac{3}{2} = x - \frac{3}{2} + \frac{3}{2} = x$$

The composite functions of f and g and of g and f both equal x, so we can assume they're inverse functions of each other; $f(x)$ is $g^{-1}(x)$, and $g(x)$ is $f^{-1}(x)$.

Example 16:

Are the functions $f(x) = 2x - 3$ and $h(x) = 3 - 2x$ inverse functions?

$$f(h(x)) = f(3 - 2x) = 2(3 - 2x) - 3 = 6 - 4x - 3 = -4x + 3.$$

They are not inverse functions because $f(h(x)) \neq x$.

SELF-TEST 6:

Are the given functions inverses of each other?

1. $f(x) = 4x$ $g(x) = \frac{x}{4}$

2. $f(x) = \frac{7}{8}x$ $h(x) = \frac{8}{7}x$

3. $f(x) = \frac{(1-x)}{x}$ $j(x) = \frac{1}{x+1}$

4. $f(x) = \sqrt[3]{x+4}$ $k(x) = x^3 + 4$

5. $f(x) = x + 2$ $r(x) = 2 - x$

ANSWERS:

1. $f(g(x)) = f\left(\frac{x}{4}\right) = 4\left(\frac{x}{4}\right) = x$ $g(f(x)) = g(4x) = \frac{4x}{4} = x$

These are inverse functions.

2. $f(h(x)) = f\left(\frac{8}{7}x\right) = \frac{7}{8}\left(\frac{8}{7}x\right) = x$ $h(f(x)) = h\left(\frac{7}{8}x\right) = \frac{8}{7}\left(\frac{7}{8}x\right) = x$

These are inverse functions.

3. $f(j(x)) = f\left(\dfrac{1}{x+1}\right) = \dfrac{1 - \dfrac{1}{x+1}}{\dfrac{1}{x+1}} = \dfrac{x+1-1}{x+1} \cdot \dfrac{x+1}{1} = x$

$j(f(x)) = j\left(\dfrac{1-x}{x}\right) = \dfrac{1}{\dfrac{1-x}{x}+1} = \dfrac{1}{\dfrac{1-x+x}{x}} = \dfrac{1}{\dfrac{1}{x}} = x$

These are inverse functions.

4. $f(k(x)) = f(x^3 + 4) = \sqrt[3]{x^3 + 4 + 4} = \sqrt[3]{x^3 + 8}$, which is not x, so these are not inverse functions.

5. $f(r(x)) = f(2 - x) = 2 - x + 2 = -x$, which is not x, so these are not inverse functions.

3 Graphs of Functions

In chapter 2 we learned about functions and their domains and ranges. One way we found the domain and range of a function was by looking at its graph. We found the domain by determining how far to the right or the left the graph could go, and if there were any x values left out of the graph. We found the range by determining how far up and down the graph could go, and if there were any y values left out of the graph. In chapter 2 we gave you the graph. In this chapter you'll learn how to sketch the graph of a given function, and we'll expect you to do this without the help of a graphics calculator. We'll use graphics calculators only to check our work, not to do it for us. If you're not already used to using a graphics calculator, don't panic. We'll walk you through how to use a TI (Texas Instruments)-89 graphics calculator to check your work. In this chapter we'll learn all about the graphing techniques used to graph a variety of polynomial and rational functions and we'll also review how to find intercepts, slope, and how to write equations of lines. The word "graph" actually means "picture." So get ready to draw pictures of all those functions we worked with in chapter 2. If you don't have a package of graph paper, you better run out right now and buy one (two would be even better). Don't worry, we'll wait right here for you. Okay, have you got that graph paper? Then we're ready to rock and roll, as we used to say back in the last millennium. Before we start graphing, let's review some of the concepts essential to graphing.

1 Intercepts

For all of the graphs you will sketch here, we expect you to label all the intercepts. Intercepts are some of the most important points on a graph. In the following figure we'd like you to pay close attention to the coordinates of the points we've labeled.

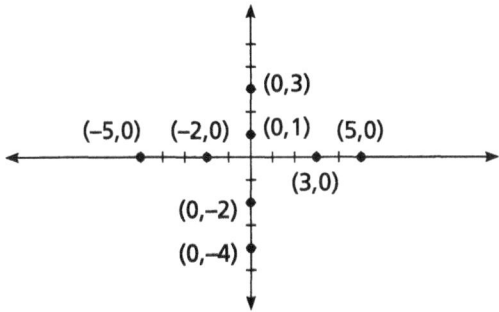

All the points on the x axis are called x-intercepts. All the points on the y axis are called y-intercepts. Notice that *every x-intercept has a y value of 0*. By the same token, *every y-intercept has an x value of 0*. We need to keep these two important facts in mind when we're trying to find our intercepts.

To find the x-intercept, substitute 0 for y and solve for x.
To find the y-intercept, substitute 0 for x and solve for y.

Example 1:

Given $y = 2x + 4$, find the x and y intercepts.

Solution:

First let's find the x-intercept by substituting 0 for y and then solving for x.

$y = 2x + 4$
$0 = 2x + 4$
$-4 = 2x$
$-2 = x$ The x-intercept is the point $(-2, 0)$. This is the point where the graph crosses the x axis.

Now let's find the y-intercept by substituting 0 for x, then solving for y.

$y = 2x + 4$
$y = 2(0) + 4$
$y = 4$ The y-intercept is the point $(0,4)$. We'll sketch the graph of this function in the next section.

Example 2:
Find the intercept(s) of the function $y = x^3 - x$.

To find the y-intercept we'll let $x = 0$, then solve for y.

$y = x^3 - x$
$y = (0)^3 - 0 = 0$ Our y-intercept is $(0,0)$.

To find the x-intercept, we'll let $y = 0$, then solve for x.

$y = x^3 - x$
$0 = x^3 - x$ Factor out the greatest common factor of x (see chapter 1, "Factoring," page 17).
$0 = x(x^2 - 1)$ Now apply the difference of two squares.
$0 = x(x - 1)(x + 1)$ Set each of the factors equal to 0 and solve for x.
$x = 0, x - 1 = 0, x + 1 = 0$
$x = 0, x = 1, x = -1$ The x-intercepts are $(0,0)$, $(1,0)$, $(-1,0)$.

We will sketch the graph of this function in "Graphs of Linear Functions" later in this chapter.

Example 3:
Find the intercepts of the graph of $y = 5x^2 + 2x - 4$.

We'll find the y-intercepts by letting $x = 0$ and solving for y.

$y = 5x^2 + 2x - 4$
$y = 5(0)^2 + 2(0) - 4 = -4$ The y-intercept is $(0,-4)$.

We'll find the x-intercepts by letting $y = 0$ and solving for x.

$y = 5x^2 + 2x - 4$
$0 = 5x^2 + 2x - 4$
$a = 5, b = 2, c = -4$

This is a quadratic equation (second degree—highest exponent is 2) that doesn't factor, so we'll use the quadratic formula

$$x = \frac{-b + \sqrt{b^2 - 4ac}}{2a}.$$

$$x = \frac{-2 + \sqrt{2^2 - 4(5)(-4)}}{2(5)} = \frac{-2 + \sqrt{4 + 80}}{10} = \frac{-2 + \sqrt{84}}{10} = \frac{-2 + 2\sqrt{21}}{10}$$

$$= \frac{-1 + \sqrt{21}}{5} \approx .7165 \text{ or } \frac{-1 - \sqrt{21}}{5} \approx -1.117$$

The x-intercepts are $\left(\frac{-1 + \sqrt{21}}{5}, 0\right), \left(\frac{-1 - \sqrt{21}}{5}, 0\right)$.

Example 4:

Find the intercepts of $y = \frac{x^2 - 36}{x^3 + x^2 + 1}$.

Solution:

Let's find the y-intercept by substituting 0 for x and solving for y.

$y = \frac{0^2 - 36}{0^3 + 0^2 + 1} = -36$ The y-intercept is the point (0, –36).

We'll find the x-intercept by substituting 0 for y and solving for x. We know that the only way a fraction can equal 0 is if the numerator equals 0, so we'll set the numerator equal to 0 and solve for y.

$x^2 - 36 = 0$ Factor using the difference of two squares.
$(x + 6)(x - 6) = 0$ Set each factor equal to 0 and solve for x.
$x + 6 = 0, x - 6 = 0$
$x = -6, x = 6$ The x-intercepts are (–6,0) and (6,0).

SELF-TEST 1:

Find the intercepts for each of the following functions:

1. $y = 3x + 4$
2. $y = -2x + 1$
3. $y = x^2 - 4x + 3$
4. $y = 2x^2 - x - 1$
5. $y = x^4 - 5x^2 + 4$
6. $y = |x + 1|$
7. $y = \frac{x - 1}{x + 4}$
8. $y = \frac{10}{x + 5}$

ANSWERS:

1. x-intercept: $\left(-\frac{4}{3}, 0\right)$ y-intercept: $(0,4)$

$y = 3x + 4$
$0 = 3x + 4$
$-4 = 3x$
$x = -\frac{4}{3}$

$y = 3x + 4$
$y = 3(0) + 4 = 4$

2. x-intercept: $\left(\frac{1}{2}, 0\right)$ y-intercept: $(0,1)$

$y = -2x + 1$
$0 = -2x + 1$
$-1 = -2x$
$x = \frac{1}{2}$

$y = -2x + 1$
$y = -2(0) + 1 = 1$

3. x-intercepts: $(1,0), (3,0)$ y-intercept: $(0,3)$

$y = x^2 - 4x + 3$
$0 = x^2 - 4x + 3$
$0 = (x - 1)(x - 3)$
$x - 1 = 0, x - 3 = 0$
$x = 1, x = 3$

$y = x^2 - 4x + 3$
$y = 0^2 - 4(0) + 3 = 3$

4. x-intercepts: $\left(-\frac{1}{2}, 0\right), (1,0)$ y-intercept: $(0,-1)$

$y = 2x^2 - x - 1$
$0 = (2x + 1)(x - 1)$
$2x + 1 = 0, x - 1 = 0$
$2x = -1, x = 1$
$x = -\frac{1}{2}$

$y = 2x^2 - x - 1$
$y = 2(0)^2 - 0 + 1 = -1$

5. x-intercepts: $(2,0), (-2,0), (1,0), (-1,0)$ y-intercept: $(0,4)$

$y = x^4 - 5x^2 + 4$
$0 = (x^2 - 4)(x^2 - 1)$
$0 = (x - 2)(x + 2)(x - 1)(x + 1)$
$x = 2, x = -2, x = 1, x = -1$

$y = x^4 - 5x^2 + 4$
$y = 0^4 - 5(0)^2 + 4 = 4$

6. x-intercept: $(-1,0)$ y-intercept: $(0,1)$

$y = |x + 1|$
$0 = x + 1$
$x = -1$

$y = |x + 1|$
$y = |0 + 1| = 1$

7. x-intercept: $(1,0)$ y-intercept: $\left(0, -\frac{1}{4}\right)$

$y = \frac{x - 1}{x + 4}$

$0 = x - 1$
$x = 1$

$y = \frac{x - 1}{x + 4}$

$y = \frac{0 - 1}{0 + 4} = -\frac{1}{4}$

8. x-intercept: none (this graph does not cross the x-axis)

This is a fraction, so the numerator has to equal 0 for the fraction to equal 0, but because the numerator doesn't have a variable in it the numerator will always be 10, never 0. Therefore there isn't any x-intercept.

y-intercept: (0,2)

$y = \dfrac{10}{x+5}$

$y = \dfrac{10}{0+5} = 2$

2. Slope of a Straight Line

The slope of a line is a measure of how steeply a line rises or falls. It's actually a ratio of the vertical change to the horizontal change between two points on the graph. The vertical change is the change in the y coordinates between two points on the graph, and the horizontal change is the change in the x coordinates between two points on the graph. We'll refer to the two points as (x_2, y_2) and (x_1, y_1). In mathematical notation, the Greek letter Δ (delta) means "change." The symbol for slope is m.

$$m = \dfrac{\text{vertical change}}{\text{horizontal change}} = \dfrac{\Delta y}{\Delta x} = \dfrac{(y_2 - y_1)}{(x_2 - x_1)}$$

When we use our slope formula it doesn't make any difference which one we call the first point and which one we call the second point; we'll still get the same slope when we use the slope formula $m = \dfrac{(y_2 - y_1)}{(x_2 - x_1)}$. Four graphs are shown below. The line in (a) rises up, in (b) it falls, (c) is a horizontal line, and (d) is a vertical line. Let's calculate the slopes for all four graphs using the slope formula.

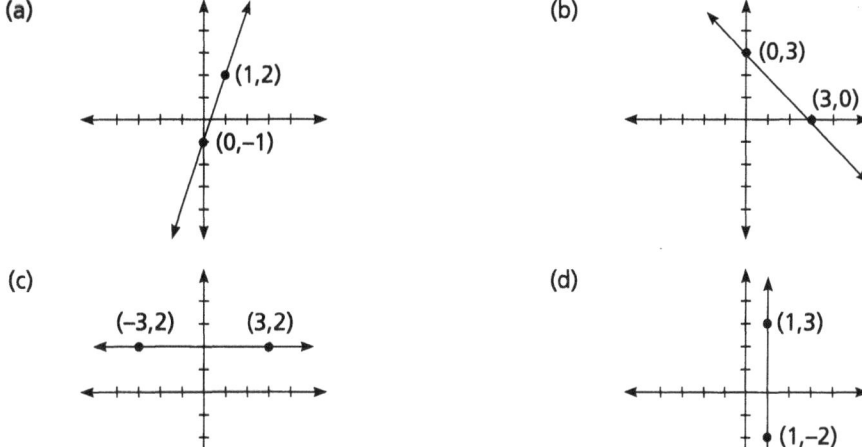

The slope of the line in figure (a) on page 57 is $m = \frac{(y_2 - y_1)}{(x_2 - x_1)} = \frac{(2 - -1)}{(1 - 0)} = \frac{3}{1} = 3$.

This tells us that if we wanted to move from one point on the line to another point on the line, all we would have to do is move up three units vertically and one unit to the right horizontally.

The slope of a line that rises is always positive.

The slope of the line in figure (b) is $m = \frac{(3 - 0)}{(0 - 3)} = \frac{3}{-3} = -1$.

The slope of a line that falls is always negative.

The slope of the line in figure (c) is $m = \frac{(2 - 2)}{(-3 - 3)} = \frac{0}{-6} = 0$.

The slope of a horizontal line is always 0.

The slope of the line in figure (d) is $m = \frac{(3 - -2)}{(1 - 1)} = \frac{5}{0} =$ undefined.

The slope of a vertical line is undefined.

SELF-TEST 2: Calculate the slopes of the following graphs:

1.
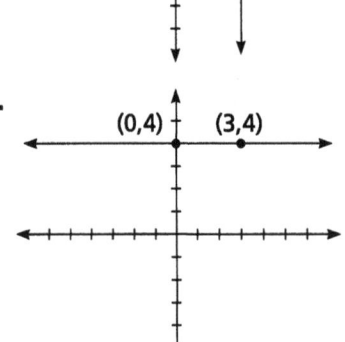

2.
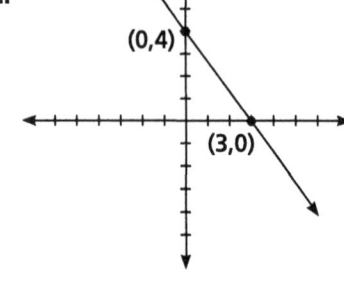

3.

4.
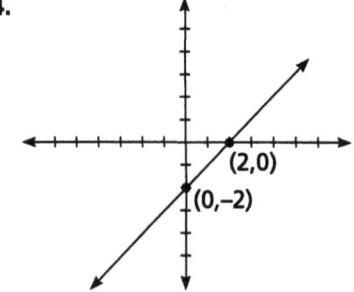

ANSWERS:

$$m = \frac{(y_2 - y_1)}{(x_2 - x_1)}$$

1. $m = \frac{(4-0)}{(3-3)} = \frac{4}{0}$ = undefined
The slope of a vertical line is undefined.

2. $m = \frac{(4-0)}{(0-3)} = -\frac{4}{3}$
The slope of a line that falls is negative.

3. $m = \frac{(4-4)}{(3-0)} = 0$
The slope of a horizontal line is 0.

4. $m = \frac{(0--2)}{(2-0)} = \frac{2}{2} = 1$
The slope of a line that rises is positive.

3 Writing the Equation of a Straight Line

There are two basic formulas we can use to write the equation of a line. The first is $y = mx + b$, where b is the y-intercept and m is the slope. We can't use this formula unless we know the y-intercept. Remember, we can tell a point is a y-intercept if its x coordinate is 0. The second formula is $y - y_1 = m(x - x_1)$; we don't need to know the y-intercept to use this formula.

Example 5:

Write the equation of the line with $m = 2$ that passes through the point (0,5).

Solution:

We know the point (0,5) is the y-intercept because x is 0; $b = 5$. If we substitute $b = 5$ and $m = 2$ into the formula $y = mx + b$ we get the equation of the line, which is $y = 2x + 5$.

Example 6:

Write the equation of the line with $m = 3$ that passes through the point (2,5).

Solution:

We know (2,5) is not the y-intercept because x is 2, not 0. Since we don't know the y-intercept we have to use the formula $y - y_1 = m(x - x_1)$; $m = 3$, $x_1 = 2$, $y_1 = 5$.

$y - y_1 = m(x - x_1)$
$y - 5 = 3(x - 2)$
$y - 5 = 3x - 6$
$y = 3x - 1$

We know the y-intercept for this line is (0,–1) because it's in $y = mx + b$ form. When a line is in this form, the slope is the coefficient of x.

Example 7:

Write the equation of the line that passes through the points (2,4) and (−1,−5).

Solution:

First, we have to decide which formula to use. Neither of the points we were given has an x value of 0, so we know they're not the y-intercept. We'll use $y - y_1 = m(x - x_1)$. Before we use this formula we have to find m, the slope. Let's call (2,4) our first point and (−1,−5) our second point.

$$m = \frac{(y_2 - y_1)}{(x_2 - x_1)} = \frac{(-5 - 4)}{(-1 - 2)} = \frac{-9}{-3} = 3$$

$$y - y_1 = m(x - x_1)$$
$$y - 4 = 3(x - 2)$$
$$y - 4 = 3x - 6$$
$$y = 3x - 2$$

Example 8:

Write the equation of the line that passes through the points (0,4) and (3,4).

Solution:

Before we write the equation of the line, notice that both points have the same y value, 4. The graph of the line that passes through the points (0,4) and (3,4) is clearly a horizontal line. Every point on this line has a y value of 4. We know the slope of any horizontal line is 0. If we used the slope formula we would get a 0. One way to find the equation of the line is to use one of the two formulas we have, to write the equation of a line. We can use the first one because we have a y-intercept of (0,4); $b = 4$, $m = 0$.

$$y = mx + b$$
$$y = 0(x) + 4 = 4$$
$$y = 4$$

Another way to write the equation of the horizontal line that passes through the given points is to know that the equation of any horizontal line will always be $y =$ a constant.

You will not see an x in the equation of a horizontal line.

Example 9:

Write the equation of the line that passes through the points (4,0) and (4,3).

Solution:

Before we write the equation of the line, notice that both points have the same x value, 4. The graph of the line that passes through the points (4,0) and (4,3) is clearly a vertical line. Every point on this line has an x value of 4. We know that the slope of any vertical line is undefined. If we used the slope formula our answer would be undefined. The easiest way to write the equation of any vertical line is to know that the equation of any vertical line is $x =$ a constant.

In this case the equation of this line is $x = 4$.

SELF-TEST 3:

Write the equation of the line that:

1. passes through the point (0,7) with a slope of –2

2. passes through the points (2,4) and (–2,–4)

3. passes through the point (–5,–1) with a slope of 3

4. has an x-intercept of 6 and a y-intercept of –3

5. passes through the points (0,1) and (2,1)

6. passes through the points (5,6) and (5,0)

ANSWERS:

1. Since $x = 0$, we know our y-intercept is 7; $b = 7$, and $m = -2$. We can use the first formula:
$y = mx + b$
$y = -2x + 7$

2. First let's start by finding the slope $m = \frac{(y_2 - y_1)}{(x_2 - x_1)} = \frac{(-4 - 4)}{(-2 - 2)} = \frac{-8}{-4} = 2$.

    We don't know the y-intercept, so we have to use the second formula:
$y - y_1 = m(x - x_1)$
$y - 4 = 2(x - 2)$
$y - 4 = 2x - 4$
$y = 2x$

3. We don't know the y-intercept, so we'll use the second formula:
$y - y_1 = m(x - x_1)$
$y - -1 = 3(x - -5)$
$y + 1 = 3x + 15$
$y = 3x + 14$

4. The points we're given are (6,0) and (0,–3). We need to find the slope.

$m = \frac{(y_2 - y_1)}{(x_2 - x_1)} = \frac{(-3-0)}{(0-6)} = \frac{1}{2}$. We know the y-intercept, $b = -3$.

$y = mx + b$

$y = \frac{1}{2}x - 3$

5. We know this is a horizontal line because both points have the same y value. We also know that the equation of any horizontal line is y = a constant. y = 1 is the equation of this horizontal line.

6. We know this is a vertical line because both points have the same x value. We also know that the equation of any vertical line is x = a constant. x = 5 is the equation of this vertical line.

4 Graphs of Linear Functions

A linear equation is a polynomial equation whose highest exponent is 1. It's also called a first-degree equation. The graph of a linear equation is always a straight line. We need only two points to graph a straight line. The points we want to see on your graphs are the intercepts. You'll find graphing linear equations to be very simple, especially since we reviewed how to find intercepts in section 1 of this chapter.

Example 10:
Graph $2x + 4y = 8$.

Solution:
We'll start by finding the *x* and the *y* intercepts; then we'll plot the points and connect them with a straight line. Be sure to extend the line beyond those points and put arrows at the ends of the lines to show that the graph doesn't end at these points. If you don't put arrows at the end, it's a line segment, which is only part of the graph. Write the equation of the line on your graph and label the intercepts.

x-intercept: (4,0)   y-intercept: (0,2)
2x + 4(0) = 8   2(0) + 4y = 8
    2x = 8   4y = 8
     x = 4   y = 2

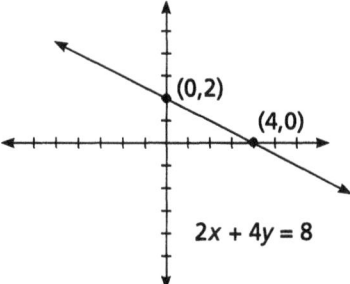

Example 11:

Graph $-2x + 4y = 8$.

Solution:

Notice that this equation is almost identical to the equation in example 10. The only difference is the negative in front of the $2x$. Let's see what kind of an effect this has on our graph.

x-intercept: (–4,0)   y-intercept: (0,2)
$-2x + 4(0) = 8$ $-2(0) + 4y = 8$
$\quad\quad -2x = 8$ $\quad\quad 4y = 8$
$\quad\quad\quad x = -4$ $\quad\quad\quad y = 2$

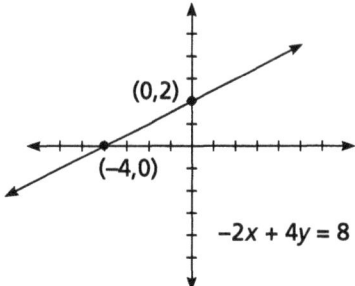

The negative in front of the $2x$ reverses the line's direction. When we write the equation $-2x + 4y = 8$ in $y = mx + b$ form, it's $y = \frac{1}{2}x + 2$. Now it's rising instead of falling. The slope of the line in example 9 is positive $\frac{1}{2}$ and its y-intercept is 2. When we write the equation $2x + 4y = 8$ in $y = mx + b$ form it's $y = -\frac{1}{2}x + 2$. The slope of the line in example 8 is negative $\frac{1}{2}$ and its y-intercept is 2.

Because the slope is negative, this line falls.

Example 12:

Graph $y = 2$.

Solution:

The first thing we notice here is that the equation doesn't have an x in it. It's the horizontal line that has a y value of 2 for every point on the line. Therefore it has a y-intercept of $(0,2)$ and no x-intercept.

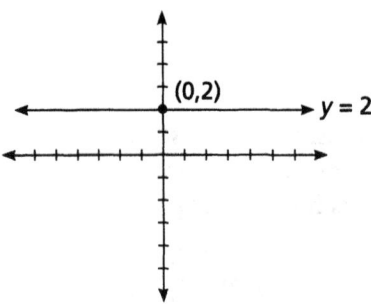

Example 13:

Graph $x = 2$.

Solution:

The first thing we notice here is that the equation doesn't have a y in it. It's the vertical line that has an x value of 2 for every point on the line. Therefore it has an x-intercept of $(2,0)$ and no y-intercept.

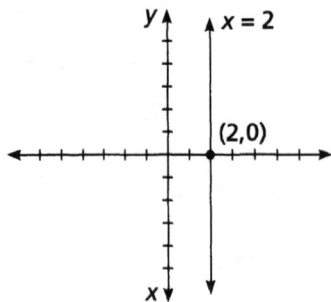

Graphs of Functions

Example 14:

Graph $y = x$.

Solution:

This equation has an x and a y, so we know it's not a horizontal or a vertical line. The equation doesn't have a constant in it. When an equation doesn't have a constant in it, it only has one intercept, the origin, (0,0). We need two points to graph a line, so in this special case we'll arbitrarily choose a value for x and solve for y to get a second point on the line. Suppose we pick a value of 2 for x; then y would equal 2. This gives us our second point, (2,2). Now we can graph the line.

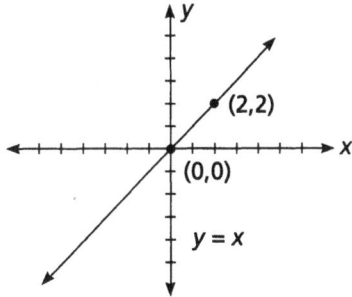

SELF-TEST 4:

Graph the line for each of the following equations. Label intercepts.

1. $3x + 6y = 12$
2. $-3x + 6y = 12$
3. $y = 1$
4. $x = -1$
5. $5x = -2y$

ANSWERS:

1. x-intercept: (4,0)　　　y-intercept: (0,2)
$3x + 6y = 12$　　　　　$3x + 6y = 12$
$3x + 6(0) = 12$　　　　$3(0) + 6y = 12$
$3x = 12$　　　　　　　$6y = 12$
$x = 4$　　　　　　　　$y = 2$

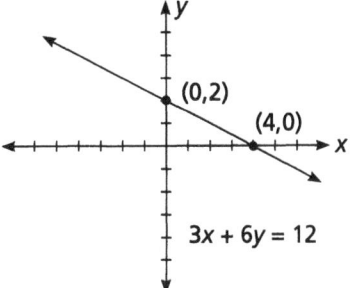

2. x-intercept: (−4,0)   y-intercept: (0,2)
$-3x + 6y = 12$ $-3x + 6y = 12$
$-3x + 6(0) = 12$ $-3(0) + 6y = 12$
$-3x = 12$ $6y = 12$
$x = -4$ $y = 2$

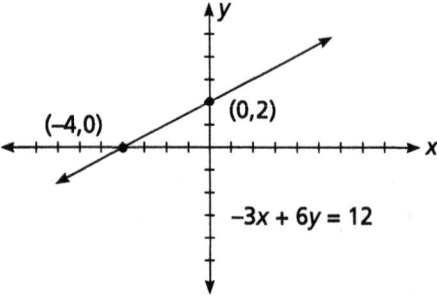

3. Since there is a y but no x, we know this is a horizontal line that has a y value of 1 for every point on the line. Therefore it has a y-intercept of (0,1) and no x-intercept.

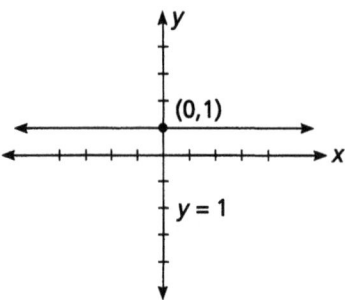

4. Since there is an x but no y, we know this is a vertical line that has an x value of −1 for every point on the line. Therefore it has an x-intercept of (−1,0) and no y-intercept.

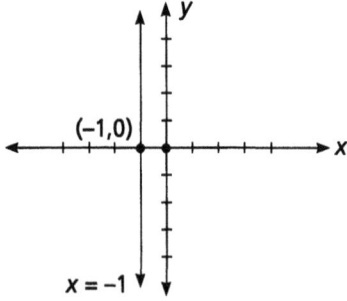

5. Notice that there is an x and a y but no constant. That tells us there is only one intercept, (0,0). For this special case we arbitrarily pick a value for x and solve for its corresponding y value. Let's pick a value of 1 for x.

$5x = -2y$

$5(1) = -2y$

$y = -\frac{5}{2}$ $\left(1, -\frac{5}{2}\right)$

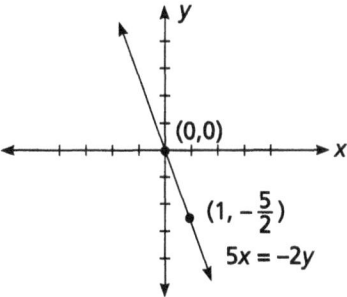

5 Graphs of Quadratic Functions

A quadratic equation is a second-degree equation (highest exponent is 2) of the form $f(x) = ax^2 + bx + c$. Its graph is always a parabola that is symmetric around an axis, called its axis of symmetry. The axis of symmetry is the vertical line that passes through the x coordinate of its vertex. The vertex of a parabola is its highest or lowest point, or the peak of the graph. To graph a parabola we need to know whether it opens up or down, its intercepts, and its axis of symmetry. We need to keep in mind that a is the coefficient of the square, b is the coefficient of the first-power term, and c is the constant.

Parabolas That Open Up and That Open Down

If $a > 0$, the parabola opens up, like the left figure on page 68. Shown is the graph of the standard quadratic functions, $y = x^2$ and $y = -x^2$. In the first function $a = 1$; in the second, $a = -1$. If $a < 0$, the parabola opens down, just like the figure to the right. Would the graph of the equation $f(x) = 4x^2 + 3x + 1$ be a parabola that opens up or down? It would open up because $a = 4$ and $4 > 0$. Would the graph of $f(x) = -2x^2 - 4x - 1$ be a parabola that opens up or down? It would open down because $a = -2$, and $-2 < 0$. So as soon as we spot a minus sign in front of the first term of the quadratic, we know the parabola opens down. If there is no minus sign, then the parabola opens up.

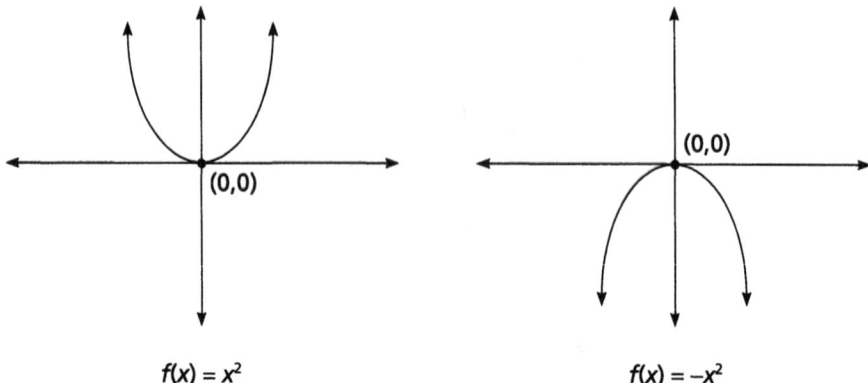

$f(x) = x^2$ $f(x) = -x^2$

The Coordinates of the Vertex

The vertex of a parabola is the highest point on a graph that opens down and the lowest point on a graph that opens up. One way to find the vertex of a parabola is to use the formula V: $\left(\frac{-b}{2a}, f\left(\frac{-b}{2a}\right)\right)$. The x coordinate of its vertex gives us its axis of symmetry. The axis of symmetry is the vertical line that passes through the x-coordinate of the vertex. Its equation is $x = \frac{-b}{2a}$.

Example 15:

Find the vertex for the quadratic function $f(x) = 2x^2 + 8x + 7$.

Solution:

The x-coordinate for the vertex of the function is $\frac{-b}{2a} = \frac{-8}{2(2)} = -2$.

The y-coordinate for the vertex of the function is $f\left(\frac{-b}{2a}\right) = f(-2) = 2(-2)^2 + 8(-2) + 7 = -1$.

The vertex is $(-2, -1)$. Its axis of symmetry is $x = -2$.

Example 16:

Find the vertex and the intercepts for the quadratic function. Does it open up or down? Sketch the graph; $f(x) = -x^2 + 6x - 8$.

Graphs of Functions

Solution:

The x-coordinate of the vertex is $\dfrac{-b}{2a} = \dfrac{-6}{2(-1)} = \dfrac{-6}{-2} = 3$.

The y-coordinate of the vertex is $y = f(3) = -(3)^2 + 6(3) - 8 = 1$.
Vertex: $(3,1)$. Its axis of symmetry is $x = 3$.
The intercepts are:

x-intercepts: $(4,0), (2,0)$ $\quad$ y-intercept: $(0,-8)$
$y = -x^2 + 6x - 8$ $\qquad\qquad$ $y = -x^2 + 6x - 8$
$0 = -(x^2 - 6x + 8)$ $\qquad\quad$ $y = -(0)^2 + 6(0) - 8$
$0 = -(x-2)(x-4)$ $\qquad\qquad$ $y = -8$
$x - 2 = 0, x - 4 = 0$
$x = 2, x = 4$

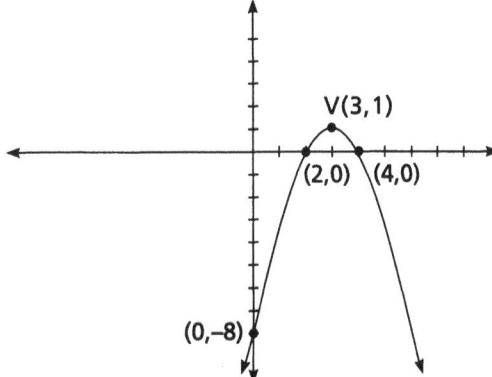

We know this graph opens down because a is -1, which is less than 0.

SELF-TEST 5: Find the vertex, intercepts, and sketch the graphs of the following functions:

1. $f(x) = -2x^2 + 4x$
2. $f(x) = x^2 + 2x - 1$
3. $f(x) = x^2 - 3x - 4$
4. $f(x) = -x^2 + 1$

ANSWERS:

1. $f(x) = -2x^2 + 4x$
   Vertex: $(1,2)$
   $x = -\dfrac{b}{2a} = -\dfrac{-4}{2(-2)} = 1$ $\qquad$ $y = f(1) = -2(1)^2 + 4(1) = 2$
   x-intercepts: $(0,0), (2,0)$ $\qquad$ y-intercept: $(0,0)$
   $0 = -2x^2 + 4x$ $\qquad\qquad\qquad$ $y = -2(0)^2 + 4(0) = 0$
   $0 = -2x(x - 2)$
   $2x = 0, x - 2 = 0$
   $x = 0, x = 2$
   $a < 0$, so this parabola opens down.

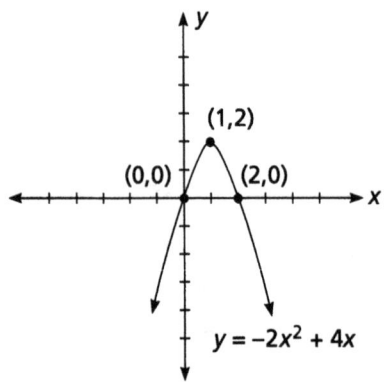

$y = -2x^2 + 4x$

2. $f(x) = x^2 + 2x - 1$
Vertex: $(-1, -2)$

$x = \dfrac{-b}{2a} = \dfrac{-2}{2(1)} = -1$

$y = f(-1) = (-1)^2 + 2(-1) - 1 = -2$

x-intercepts: $(-1 + \sqrt{2}, 0), (-1 - \sqrt{2}, 0)$
$y = x^2 + 2x - 1$ doesn't factor, so we'll use the quadratic formula.

y-intercept: $(0, -1)$
$y = (0)^2 + 2(0) - 1 = -1$

$x = \dfrac{-b \pm \sqrt{b^2 - 4ac}}{2a} = \dfrac{-2 \pm \sqrt{2^2 - 4(1)(-1)}}{2(1)} = \dfrac{-2 \pm \sqrt{4+4}}{2} = \dfrac{-2 \pm \sqrt{8}}{2} = \dfrac{-2 \pm 2\sqrt{2}}{2}$

$= -1 \pm \sqrt{2} \begin{cases} -1 + \sqrt{2} \approx .414 \\ -1 - \sqrt{2} \approx -2.41 \end{cases}$

$a > 0$, so this parabola opens up.

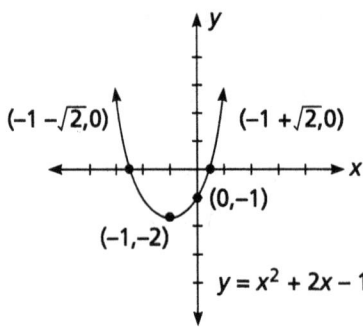

3. $f(x) = x^2 - 3x - 4$

Vertex: $\left(\dfrac{3}{2}, -\dfrac{25}{4}\right)$

$x = -\dfrac{b}{2a} = -\dfrac{-3}{2(1)} = \dfrac{3}{2}$

$y = \left(\dfrac{3}{2}\right)^2 - 3\left(\dfrac{3}{2}\right) - 4 = \dfrac{9}{4} - \dfrac{9}{2} - 4 = \dfrac{9}{4} - \dfrac{18}{4} - \dfrac{16}{4} = -\dfrac{25}{4}$

x-intercepts: (4,0), (−1,0)   y-intercepts: (0,−4)
$x^2 − 3x − 4 = 0$ $f(0) = 0^2 − 3(0) − 4 = −4$
$(x − 4)(x + 1) = 0$
$x = 4, x = −1$
$a > 0$, so this parabola opens up.

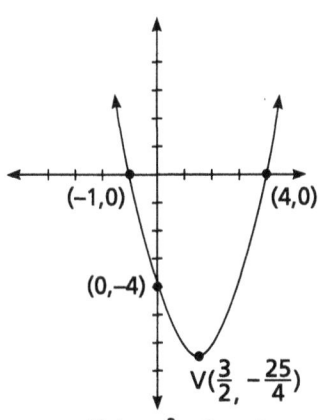

$f(x) = x^2 − 3x − 4$

4. $f(x) = −x^2 + 1$
Vertex: (0,1)

$x = \dfrac{-b}{2a} = \dfrac{0}{2(-1)} = 0$ $y = −(0)^2 + 1 = 1$

x-intercept: (1,0),(−1,0)   y-intercept: (0,1)
$0 = −x^2 + 1$ $y = −(0)^2 + 1 = 1$
$0 = −(x + 1)(x − 1)$
$x = 1, x = −1$
$a < 0$, so this parabola opens down.

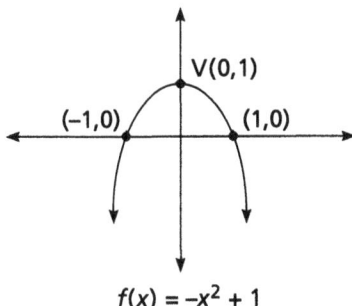

$f(x) = −x^2 + 1$

The Shifting Technique for Graphing

Now we're going to show you a technique for graphing that applies to all of the basic types of graphs we'll be working on in this book. On page 68 we showed the graph of the basic quadratic functions: $y = x^2$

and $y = -x^2$. Both graphs have only one intercept, the origin, (0,0). Both graphs are shown below.

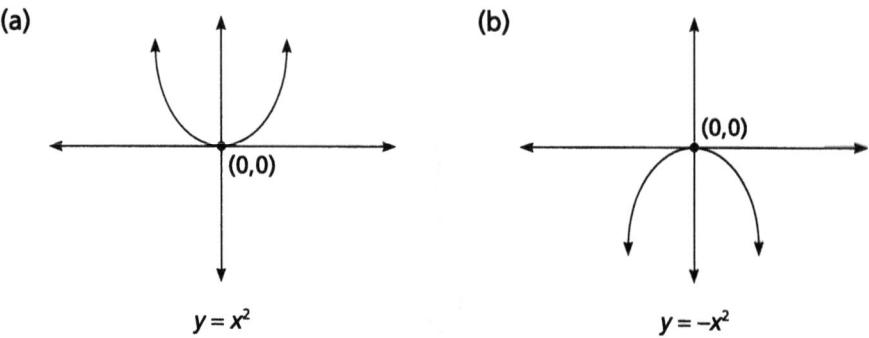

Below are the graphs of four functions. Notice what effect adding or subtracting a constant to the basic quadratic functions $y = x^2$ and $y = -x^2$ has on the graph.

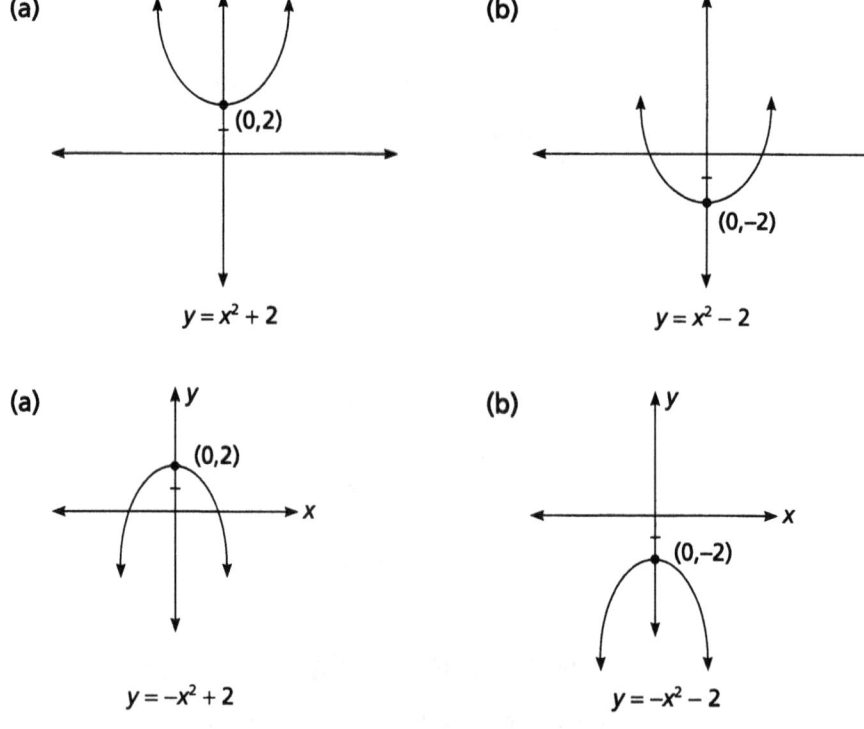

When we added a 2, the standard graph shifted up two units, and when we subtracted a 2, the standard graph moved down two units.

Graphs of Functions

Adding or subtracting a constant to a function causes a vertical shift in the basic function.

Let's see what happens if we add or subtract a constant within a function instead of adding or subtracting a constant outside the function. Look at the figures below to see what happens if we add or subtract a constant inside the basic quadratic functions. If you need to refer to the graphs of the basic quadratic functions $y = x^2$ and $y = -x^2$, they're illustrated on page 68.

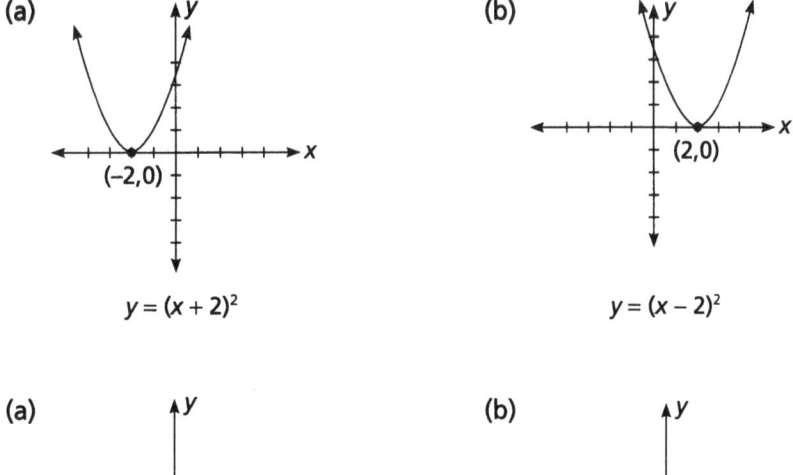

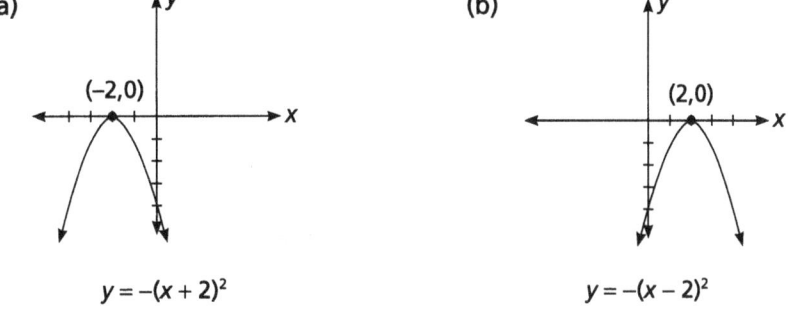

When we added a 2 within the function, the basic graph shifted two units to the left, and when we subtracted a 2 within the function, the basic graph shifted two units to the right.

Adding or subtracting a constant within the function causes a horizontal shift in the opposite direction of the basic function.

Let's see what happens when we have a vertical and a horizontal shift in the basic function.

Example 17:
Graph $(x - 3)^2 + 1$.

Solution:

We'll have to shift the basic function $y = x^2$ up one unit because a 1 was added outside the function. We'll also shift it three units to the right because a −3 was inserted inside the function.

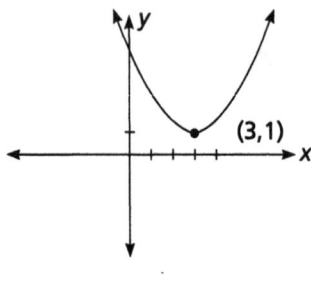

$y = (x − 3)^2 + 1$

Now that we know about the vertical and the horizontal shifts, let's see what effect multiplying the function by a constant has on the graph. The figure below shows us how multiplying by a constant greater than 1 causes the graph to compress closer to the axis of symmetry and how multiplying by a constant less than 1 causes a graph to spread out away from the axis of symmetry.

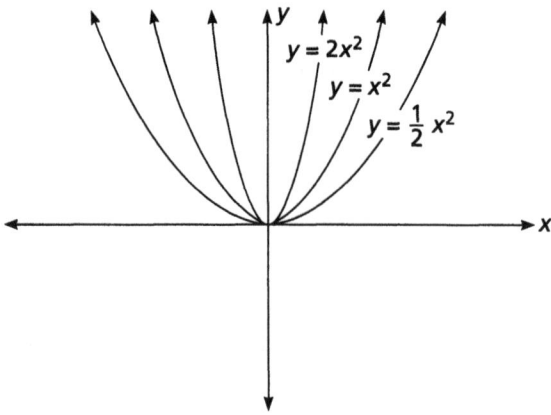

Example 18:
Graph $3(x + 2) − 4$.

Solution:

This graph will shift down four units vertically, and to the left two units horizontally. It will compress closer to the axis of symmetry.

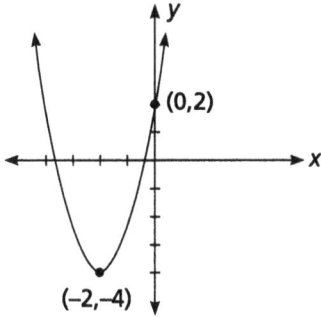

Before we give you a self-test on the shifting technique, let's go back to example 16 in this section. There we graphed $f(x) = -x^2 + 6x - 8$. When a function is written in this form it's called the general form. Another way of writing this function is $f(x) = -(x - 3)^2 + 1$. When the function is written in this form it's called the standard form. Using the shifting technique would give us the same graph as we got in example 16. For self-test 6 we're going to give you the same problems as in self-test 1. This time they'll be in standard form, so you won't have to figure out the intercepts for a new set of problems. (You can thank us later.)

SELF-TEST 6:

Describe the graphs of the following functions using the shifting technique:

1. $f(x) = -2(x - 1)^2 + 2$
2. $f(x) = (x + 1) - 2$
3. $f(x) = (x - 6) + 3$
4. $f(x) = -(x - 2)^2$
5. $f(x) = (x + 2)^2 + 2$
6. $f(x) = -x^2 + 1$

ANSWERS:

The graphs are similar to those in self-test 1.

1. This function will shift up two units and to the right one unit. It will compress around the axis of symmetry.
2. This function will shift down two units and to the left one unit.
3. This function will shift up three units and to the right six units.
4. This function will shift two units to the right.
5. This function will shift two units up and two units to the left.
6. This function will shift one unit up.

Before we move on to the next section, we'd like to show you how to use the TI (Texas Instruments)-89 graphics calculator to check your work. Suppose we want to check example 16 using our calculator to graph $f(x) = -x^2 + 6x - 8$. It doesn't matter whether we use the general form or the standard form. If you press the following buttons in the order in which we give them to you, you'll get the graph: ◆ y = − x ∧ 2 + 6 x − 8 enter F2 6. We only suggest you use the calculator to check your graph, not to do the graph for you.

6 Graphs of Polynomial Functions of Degree 3 and Higher

In sections 4 and 5 of this chapter we graphed polynomial functions of degrees 1 and 2. In this section we will graph polynomial functions of degree 3 and higher. We will start out using the shifting technique on functions that are in standard form. We will divide our graphs into ones that are to an odd power and ones that are to an even power. Shown below are the graphs of the basic functions to which we will apply the shifting technique.

(a)
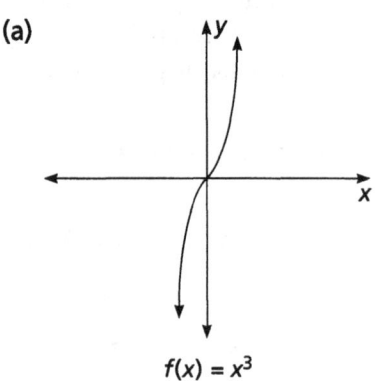
$f(x) = x^3$

(b)
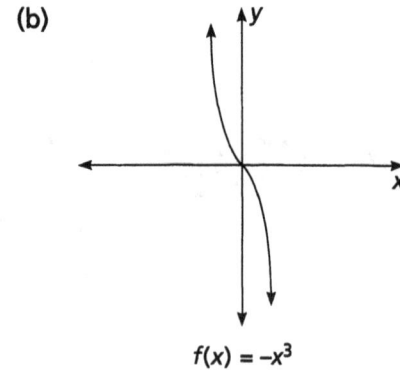
$f(x) = -x^3$

(c)
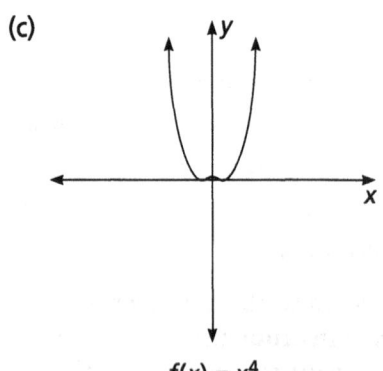
$f(x) = x^4$

(d)
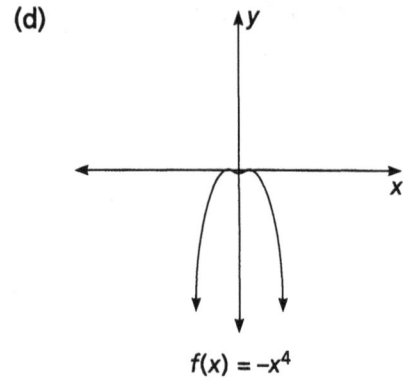
$f(x) = -x^4$

The graph of an odd-powered polynomial function with a positive leading coefficient will always be up on the right and down on the left; see figure a.

The graph of an odd-powered polynomial function with a negative leading coefficient will always be up on the left and down on the right; see figure b.

The graph of an even-powered polynomial function with a positive leading coefficient will always be up on the right and up on the left; see figure c.

The graph of an even-powered polynomial function with a negative leading coefficient will always be down on the right and down on the left; see figure d.

Example 19:

For each of the following functions, what should the form of the graph be?

a. $y = 2x^7$ **b.** $y = -2x^6$ **c.** $y = 5x^4$ **d.** $y = -4x^5$

Solutions:

a. This is an odd-powered polynomial with a positive leading coefficient, so its graph will be up on the right and down on the left.

b. This is an even-powered polynomial with a negative leading coefficient, so its graph will be down on the right and down on the left.

c. This is an even-powered polynomial with a positive leading coefficient, so its graph will be up on the right and up on the left.

d. This is an odd-powered polynomial with a negative leading coefficient, so its graph will be up on the left and down on the right.

Example 20:

Use the shifting technique to graph $f(x) = (x + 1)^3 - 2$. Label intercepts.

Solution:

This is an odd-powered polynomial with a positive leading coefficient, so its graph will be up on the right and down on the left. The 1 added inside the function will cause it to shift one unit to the left, and the negative 2 added outside the function will cause it to shift down two units. We'll find the y-intercept by substituting 0 inside the function. The y-intercept is the point (0,–2). Finding the x-intercept is more work. Another name for the x-intercept is the 0 or root of the function. To find the 0s of the function we'll use modern technology. We'll use our TI-89 graphics calculator. Just in case you're not comfortable using a calculator to find the 0s, we'll walk you through it. Press the following buttons in the order we do and you'll get the x-intercept of the function. F2 4 (x + 1) ∧ 3 − 2 , x) enter. The x-intercept is approximately (.26,0). We don't know the exact value of the x-intercept, but we can use a close approximation of it to help us get a more accurate graph. We couldn't find an exact value because we couldn't factor the given function, and there isn't a formula available for us to use on a third-degree function. Now we know the shape of the graph and the intercepts (0,−2) and ≈(.26,0).

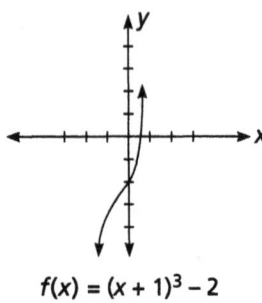

$f(x) = (x + 1)^3 - 2$

Example 21:

Use the shifting technique to graph the following functions. Label intercepts.

a. $f(x) = x^3 + 3$
b. $f(x) = (x + 3)^3$
c. $f(x) = -x^4 - 3$
d. $f(x) = -(x - 1)^5 - 2$
e. $f(x) = (x + 2)^4 - 4$

Solutions:

a. The 3 added outside the basic function will cause this graph to shift up three units.
 y-intercept: (0,3)
 x-intercept: $(\sqrt[3]{-3}, 0) \approx (-1.44, 0)$

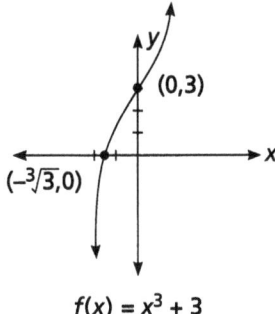

$f(x) = x^3 + 3$

b. The 3 added inside the basic function will cause this graph to shift three units to the left.
y-intercept: (0,27)
x-intercept: (−3,0)

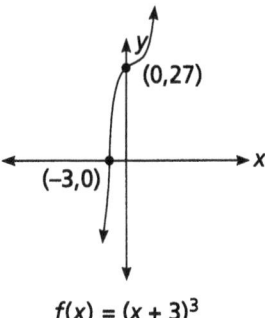

$f(x) = (x + 3)^3$

c. The negative 3 added outside the function will cause the basic function to shift down three units. The negative coefficient outside the functions causes it to open down on both ends.
y-intercept: (0,−3)
x-intercept: none

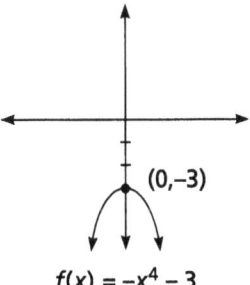

$f(x) = -x^4 - 3$

d. The negative 1 inside the basic function will cause it to shift one unit to the right. The negative 2 outside the basic function will cause it to shift two units down. The negative coefficient outside the function will cause it to be up on the left and down on the right.

y-intercept: $(0,-3)$
x-intercept: $\approx (-.15, 0)$

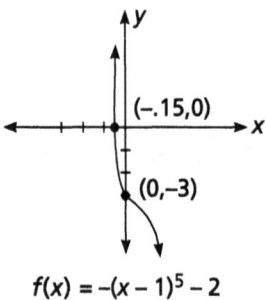

$f(x) = -(x-1)^5 - 2$

e. The 2 added inside the basic function will cause it to move two units to the left. The negative 4 outside the function will cause it to move four units down.

y-intercept: $(0, 12)$
x-intercept: $(\sqrt{2} - 2, 0), (-\sqrt{2} - 2, 0)$

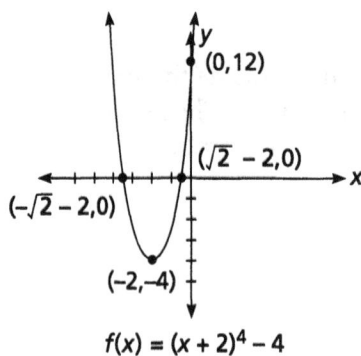

$f(x) = (x+2)^4 - 4$

SELF-TEST 7:

Graph the following functions using the shifting technique. Label intercepts.

1. $f(x) = (x-1)^4 - 2$

2. $f(x) = -(x+2)^3 - 4$

3. $f(x) = (x+1)^5 + 1$

4. $f(x) = -(x+1)^6 + 2$

Graphs of Functions

ANSWERS:

1. This graph will shift one unit to the right and two units down. It's an even-powered function with a positive leading coefficient, so it will open up on the right and up on the left.
   y-intercept: (0,−1)
   x-intercept: ≈(−.19,0), (2.19,0)

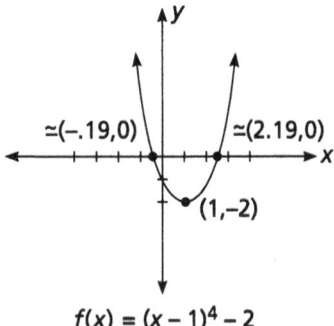

$$f(x) = (x - 1)^4 - 2$$

2. This graph is an odd-powered function that shifts two units to the left and four units down. It will be up on the left and down on the right.
   y-intercept: (0,−12)
   x-intercept: ≈(−3.6,0)

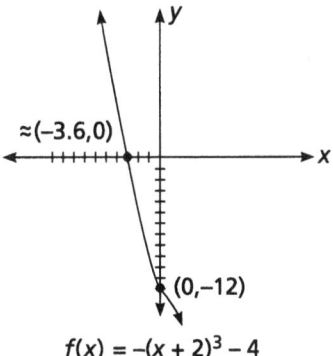

$$f(x) = -(x + 2)^3 - 4$$

3. This is an odd-powered function that shifts one unit to the left and one unit up. It will be up on the right and down on the left.
   y-intercept: (0,2)
   x-intercept: (−2,0)

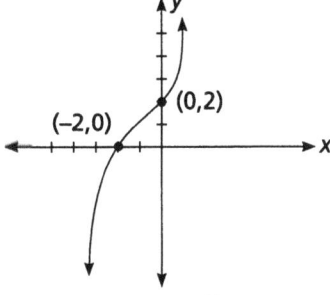

$$f(x) = (x + 1)^5 + 1$$

4. This is an even-powered function with a negative leading coefficient. It will be down on the right and down on the left. It will shift one unit to the left and two units up.
  y-intercept: (0,1)
  x-intercepts: ≈(−2.12,0), (.12,0)

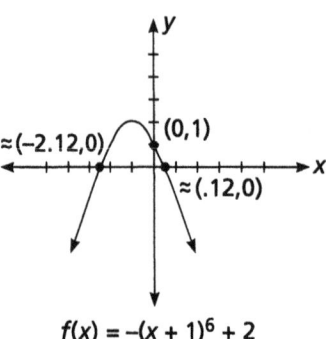

$$f(x) = -(x+1)^6 + 2$$

All the graphs we've sketched so far in this section have been in standard form. When a function is written in standard form it's easy to use the shifting technique. The following are some of the functions in example 21 written in general form. To go from standard form to general form all we did was expand the function. When the function is in general form we can't use the shifting technique, but we can still use our graphics calculator to find the x-intercepts. We can also use our general knowledge of higher-degree polynomials to find the general form of the function and to sketch the graph.

Example 21 revisited:

a. and **c.** don't expand.
b. $f(x) = x^3 + 9x^2 + 27x + 27$
d. $f(x) = -x^5 + 5x^4 - 10x^3 + 10x^2 - 5x - 1$
e. $f(x) = x^4 + 8x^3 + 24x^2 + 32x + 12$

Whether the functions are in general form or in standard form, the graphs are the same.

7 Asymptotes

What's an asymptote? An asymptote is a line the graph approaches. In this section we'll learn how to find vertical and horizontal asymptotes. We'll be working with rational functions. A rational function is the quotient of two polynomial functions. A *rational function* can be written in the form

$f(x) = \frac{p(x)}{q(x)}$ where $p(x)$ and $q(x)$ are polynomials, and $q(x) \neq 0$.

Example 22:
Shown below is the graph of the function $f(x) = \frac{1}{x}$.

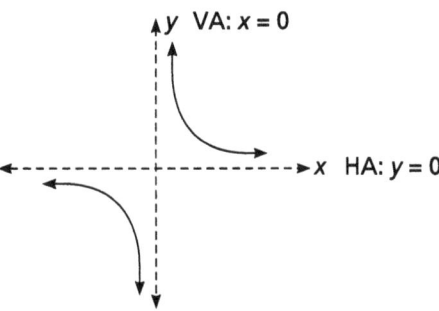

Let's find the domain of this function. Is there any value we can't allow x to be? If you said x can't be 0 because 0 in the denominator of a fraction would be undefined, you're correct. Zero is the only restriction on the domain of this function. Domain: All real numbers except $x = 0$. *A vertical asymptote results when $x = a$, where a is a restriction on the domain of a function.* The vertical asymptote is $x = 0$. Refer to table 3.1 to see what happens to the y values as the x values approach 0 from the right.

Table 3.1 Values of x and $f(x)$ in the Function $f(x) = \frac{1}{x}$ as x Values Approach 0 from the Right

x	100	10	1	$\frac{1}{10}$	$\frac{1}{100}$	$\frac{1}{1000}$
$f(x)$	$\frac{1}{100}$	$\frac{1}{10}$	1	10	100	1000

As the x values get closer to 0 from the right, the y values get larger without bound. We write that in the following way: $x \to 0^+, y \to \infty$. This would be read as follows: As x approaches 0 from the right, y approaches infinity.

Refer to table 3.2 to see what happens to the y values as the x values approach 0 from the left.

Table 3.2 Values of x and f(x) in the Function $f(x) = \frac{1}{x}$ as x Values Approach 0 from the Left

x	−100	−10	−1	$-\frac{1}{10}$	$-\frac{1}{100}$	$-\frac{1}{1000}$
f(x)	$-\frac{1}{100}$	$-\frac{1}{10}$	−1	−10	−100	−1000

As the x values approach 0 from the left, the y values approach negative infinity. We would write that as $x \to 0^-$, $y \to -\infty$. This would be read as follows: As x approaches 0 from the left, y approaches negative infinity.

Let's talk about the range of the function. Is there any y value the graph will never have? If you said 0, you're correct. We know this because the function has a 1 in the numerator with no variable, so the numerator will always be 1. One divided by anything is never 0, so the value of the fraction will never be 0. In table 3.3 notice that as the x values get larger in the positive or negative direction, the y values get smaller and approach 0.

Table 3.3 Values of x and f(x) in the Function $f(x) = \frac{1}{x}$

x	10	100	1000	10000	−10	−100
f(x)	$\frac{1}{10}$	$\frac{1}{100}$	$\frac{1}{1000}$	$\frac{1}{10000}$	$-\frac{1}{10}$	$-\frac{1}{100}$

Notice that the graph on page 83 never has a y value of 0. The graph approaches a y value of 0 as the x values approach ∞ and $-\infty$. As $x \to \infty$, $f(x) \to 0$; as $x \to -\infty$, $f(x) \to 0$. Range: All real numbers except $y = 0$. A *horizontal asymptote is of the form y = b where b is the restriction on the range of a function.*

The horizontal asymptote is $y = 0$.

Example 23:

Find the vertical asymptotes for the following functions:

a. $f(x) = \dfrac{2}{x+1}$ b. $f(x) = \dfrac{3x}{x^2 - 4}$

c. $f(x) = \dfrac{5}{x^3 + x^2 - 2x}$ d. $f(x) = \dfrac{3x}{x^4 + 8}$

Solutions:

To find vertical asymptotes we have to find the restrictions on the domain of the function. For rational functions the denominator can't equal 0 or the function would be undefined. The restrictions would be found by setting the denominator equal to 0 and solving for x.

a. V.A.: $x = -1$
$x + 1 = 0$
$x = -1$

b. V.A.: $x = 2, x = -2$
$x^2 - 4 = 0$
$(x + 2)(x - 2) = 0$
$x = -2, x = 2$

c. V.A.: $x = 0, x = 1, x = -2$
$x^3 + x^2 - 2x = 0$
$x(x^2 + x - 2) = 0$
$x(x - 1)(x + 2) = 0$
$x = 0, x = 1, x = -2$

d. V.A.: none
$x^4 + 8 = 0$
$x^4 = -8$
$x = \sqrt[4]{-8}$
$x =$ not a real number

Horizontal Asymptotes of a Rational Function

1. If the degree of the numerator is less than the degree of the denominator, the horizontal asymptote is $y = 0$.

   For example: $f(x) = \dfrac{2x^3}{3x^4 + x^3 - x}$.

   The degree of the numerator is 3. The degree of the denominator is 4. H.A.: $y = 0$.

2. If the degree of the numerator is equal to the degree of the denominator, the horizontal asymptote is the quotient of the leading coefficients. The leading coefficients are the coefficients of the terms with the highest exponents.

   For example: $f(x) = \dfrac{2x^2 + 3x - 1}{5x^2 - x + 2}$.

   The degree of both the numerator and the denominator is 2.

   H.A.: $y = \dfrac{2}{5}$.

3. If the degree of the numerator is greater than the degree of the denominator, there isn't a horizontal asymptote.

For example: $f(x) = \dfrac{4x^3 + 2x^2 - 3}{2x^2 + x + 1}$.

The degree of the numerator is 3. The degree of the denominator is 2.

H.A.: none, or we could say ∞, because the y values increase without bound.

Example 24:

Find the horizontal asymptotes of the following functions:

a. $f(x) = \dfrac{1}{x - 1}$

b. $f(x) = \dfrac{5x}{2x - 1}$

c. $f(x) = \dfrac{2x^3 + 4x - 3}{3x^2 + 4x + 2}$

d. $f(x) = \dfrac{4}{(x - 3)^4}$

Solutions:
a. The degree of the numerator is 0. The degree of the denominator is 1. The degree of the numerator is less than the degree of the denominator. H.A.: $y = 0$.
b. The degree of the numerator is 1. The degree of the denominator is 1. The degree of the numerator is equal to the degree of the denominator, so the horizontal asymptote is the quotient of the leading coefficients. H.A.: $y = \dfrac{5}{2}$.
c. The degree of the numerator is 3. The degree of the denominator is 2. The degree of the numerator is greater than the degree of the denominator. H.A.: none.
d. The degree of the numerator is 0. The degree of the denominator is 4. The degree of the numerator is less than the degree of the denominator. H.A.: $y = 0$.

Graphs of Functions

SELF-TEST 8: Find the vertical and the horizontal asymptotes of the following functions:

1. $f(x) = \dfrac{2}{x^4 - 16}$
2. $f(x) = \dfrac{1-x}{x-2}$
3. $f(x) = \dfrac{x^2 - 8}{x+2}$
4. $f(x) = \dfrac{(x+3)^2}{(x-2)^3}$
5. $f(x) = \dfrac{2x^3 + 2}{x^3 + x^2 - 2x}$
6. $f(x) = \dfrac{6x^5 - 2x^2}{x+5}$

ANSWERS:

1. V.A.: $x = 2$, $x = -2$ H.A.: $y = 0$

To find the vertical asymptotes we'll set the denominator equal to 0 and solve for x.
$x^4 - 16 = 0$
$(x^2 + 4)(x^2 - 4) = 0$ Factor using the difference of two squares.
$(x^2 + 4)(x - 2)(x + 2)$ Use the difference of two squares again.
$x = 2$, $x = -2$

To find the horizontal asymptote we'll compare the degrees of the numerator and the denominator. The degree of the numerator is 0; that of the denominator is 4. The degree of the numerator is less than the degree of the denominator, so the horizontal asymptote is $y = 0$.

2. V.A.: $x = 2$ H.A.: $y = -1$
$x - 2 = 0$ The degree of the numerator is 1, the degree of the denominator is 1.
$x = 2$ The degree of the numerator is equal to the degree of the denominator, so the horizontal asymptote is the quotient of the leading coefficients. $y = \dfrac{-1}{1} = -1$.

3. V.A.: $x = -2$ H.A.: none
$x + 2 = 0$ The degree of the numerator is 2. The degree of the denominator is 1.
$x = -2$ The degree of the numerator is greater than the degree of the denominator, so there isn't a horizontal asymptote.

4. V.A.: $x = 2$ H.A.: $y = 0$
$(x - 2)^3 = 0$ The degree of the numerator is 2. The degree of the
$x = 2$ denominator is 3. The degree of the numerator is less than the degree of the denominator, so the horizontal asymptote is $y = 0$.

5. V.A.: $x = 0$, $x = 1$, $x = -2$ H.A.: $y = 2$
$x^3 + x^2 - 2x$ The degree of the numerator is 3. The degree of the
$x(x^2 + x - 2)$ denominator is 3. The degree of the numerator is equal to
$x(x - 1)(x + 2)$ the degree of the denominator, so the horizontal
$x = 0$, $x = 1$, $x = -2$ asymptote is the quotient of the leading coefficients.
$y = \dfrac{2}{1} = 2$.

6. V.A.: $x = -5$ H.A.: none
$x + 5 = 0$ The degree of the numerator is 5. The degree of the denominator is
$x = -5$ 1. The degree of the numerator is greater than the degree of the denominator, so there isn't a horizontal asymptote.

8 Oblique Asymptotes and Graphs of Rational Functions

In this section we'll combine all our graphing skills to graph rational functions. When we do this, we'll expect you to label the intercepts and the asymptotes on all of our graphs. We'll also work on one more type of asymptote, oblique asymptotes. As we sketch our graphs, we'll use the following step-by-step procedure to graph our rational functions.

Steps for graphing rational functions:

1. Find the asymptotes.

2. Find the intercepts.

3. Figure out the shape of the graph.

Example 25:

Graph $f(x) = \dfrac{3x+3}{x-2}$.

Solution:

To find the vertical asymptote we set the denominator equal to 0 and solve for x. V.A.: $x = 2$. To find the horizontal asymptote we compare the degrees of the numerator and the denominator. The numerator and the denominator are both degree 1, so the horizontal asymptote is the quotient of the leading coefficients. H.A.: $y = 3$. To find the x-intercept let y equal 0 and solve for x. $0 = \dfrac{3x+3}{x-2}$. The only way a fraction will equal 0 is if the numerator equals 0.

$3x + 3 = 0$
$3(x + 1) = 0$
$\quad\quad x = -1$
x-intercept: $(-1, 0)$

To find the y-intercept let $x = 0$ and solve for y.

$y = \dfrac{3(0)+3}{0-2} = -\dfrac{3}{2}\quad \left(0, -\dfrac{3}{2}\right)$

Let's graph the information we have so far.

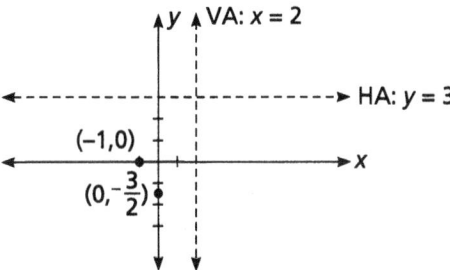

Now all we have to figure out is the form of the graph. We know the graph will approach its asymptotes. To the left of $x = 2$ we have two intercepts the graph will pass through, and we know it will approach the lines $x = 2$ and $y = 3$. To the right of $x = 2$ we know there aren't any x intercepts, so we know the graph will have to be above $y = 3$, not below it. If you have any trouble figuring out the shape of the graph, you can always substitute a few values of x into the function, which will give you some points to plot.

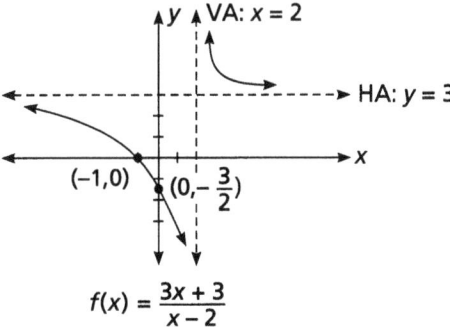

$$f(x) = \frac{3x + 3}{x - 2}$$

If you want to check your answer with a graphics calculator, press the following buttons:

◆ F1 ((3 x + 3) ÷ (x − 2)) enter ◆ F3

The graph on the screen is not as clear as the one we drew. We will only use our calculators to check the form of our graphs, not to do them for us. (And we hope you'll do the same.)

Example 26:

Graph $f(x) = \dfrac{2x}{x^2 - 4}$.

Solution:

V.A.: $x = -2, x = 2$
$x^2 - 4 = 0$
$(x + 2)(x - 2) = 0$
$x = -2, x = 2$
x-intercept: (0,0)

$2x = 0$

$x = 0$

H.A.: $y = 0$
The degree of the numerator is less than the degree of the denominator.

y-intercept: (0,0)

$f(0) = \dfrac{2(0)}{(0)^2 - 4} = 0$

Let's graph the information we have so far.

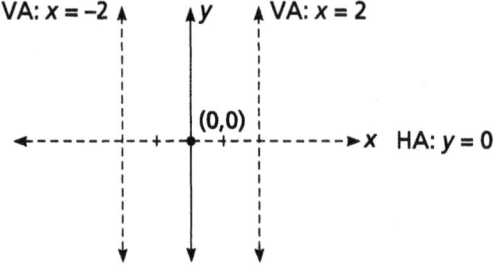

Let's look at the part of the graph that approaches $x = -2$ from the left. We know there aren't any x-intercepts here, so we know the graph has to be above or below $y = 0$. If we pick a value for x such as -3 and substitute it into the function, we get a corresponding y value that's negative. This tells us that the graph approaches the asymptotes below $y = 0$. We'll perform the same procedure for the area of the graph that approaches 2 from the right. If we pick an x value such as 3 and substitute it into the function, we get a corresponding y value that's positive. This tells us that the graph approaches the asymptotes above $y = 0$. Now we have to figure out what the graph does between -2 and 2. We

Graphs of Functions 91

know it crosses the x axis at $(0,0)$. If we pick an x value between -2 and 0 such as -1 and substitute it into the function, we get a corresponding y value that's positive. This tells us that the graph is above $y = 0$ when x is between -2 and 0. If we pick an x value between 0 and 2, such as 1, and substitute it into the function, we get a corresponding y value that's negative. This tells us that the graph is below $y = 0$ when x is between 0 and 2.

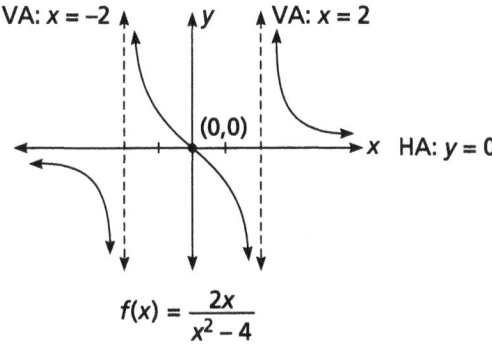

$$f(x) = \frac{2x}{x^2 - 4}$$

Example 27:

Graph $f(x) = \dfrac{2x^2}{3x^2 + 1}$.

Solution:

V.A.: none H.A.: $y = \dfrac{2}{3}$

x-intercept: $(0,0)$ y-intercept: $(0,0)$

This function has only one intercept, $(0,0)$, and only one asymptote, $y = \dfrac{2}{3}$. This tells us that the graph will approach the line $y = \dfrac{2}{3}$ and will only touch the x axis at one point $(0,0)$. This information makes it easy for us to sketch the graph. We know that all the y values will be positive because any number squared multiplied by 2 or 3 with a 1 added to it will be positive.

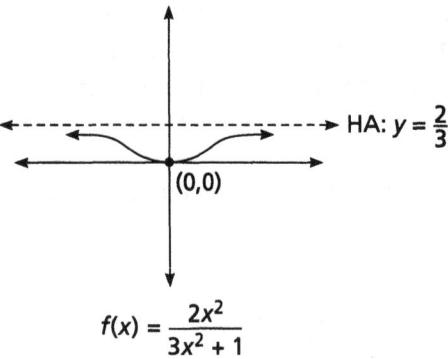

$$f(x) = \frac{2x^2}{3x^2 + 1}$$

Example 28:

Graph $f(x) = \dfrac{4x}{3x^2 + 1}$.

Solution:

V.A.: none H.A.: $y = 0$
x-intercept: (0,0) y-intercept: (0,0)

We know the graph will approach the line $y = 0$ and will touch the x axis at only one point (0,0). To find out whether the graph will be above or below the x axis, let's keep in mind that the denominator will always be positive because any number squared times 3 plus 1 will be positive. The numerator will be negative to the left of 0 and positive to the right of 0. To the left of 0 we'd have a negative divided by a positive, which would give us a negative, so the graph would be below $y = 0$ to the left of 0. To the right of 0 we'd have a positive divided by a positive, which would give us a positive, so the graph would be above $y = 0$ to the right of 0.

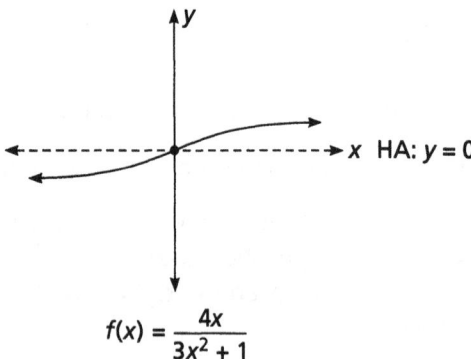

$$f(x) = \frac{4x}{3x^2 + 1}$$

Graphs of Functions 93

SELF-TEST 9: Graph the following functions. Label intercepts and asymptotes. Use your graphics calculator only to check your answers.

1. $f(x) = \dfrac{1}{x^2}$
2. $f(x) = \dfrac{2x}{x-1}$
3. $f(x) = \dfrac{2+x}{2-x}$
4. $f(x) = \dfrac{x-1}{x^2-9}$
5. $f(x) = \dfrac{x^2}{x^2+9}$
6. $f(x) = \dfrac{4(x+1)}{x(x-4)}$
7. $f(x) = -\dfrac{1}{x^2}$
8. $f(x) = \dfrac{3x}{x^2-x-2}$

ANSWERS:

1. V.A.: $x = 0$ H.A.: $y = 0$
x-intercept: none   y-intercept: none

No matter what number we substitute for x, $\dfrac{1}{x^2}$ will always be positive. This graph is very similar to that in example 27. As $x \to 0$, $y \to \infty$; as $x \to \infty$, $y \to 0$; and as $x \to -\infty$, $y \to 0$.

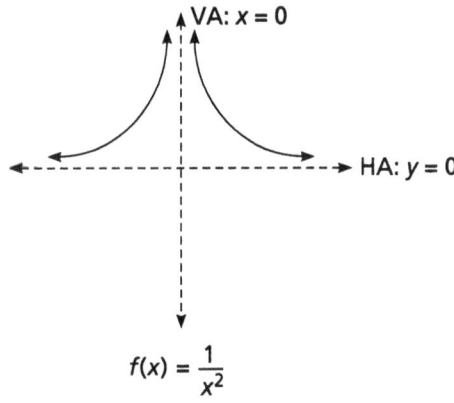

$f(x) = \dfrac{1}{x^2}$

2. V.A.: $x = 1$ H.A.: $y = 2$
x-intercept: (0,0)   y-intercept: (0,0)

If we draw what we have so far it helps us to determine the shape of the graph.

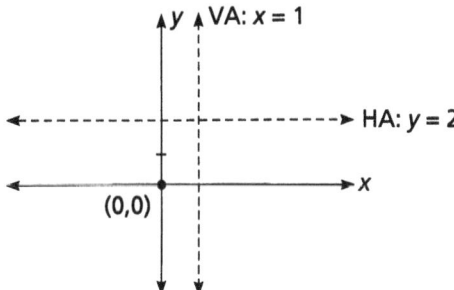

To the left of $x = 1$ we have an intercept the graph has to pass through as it approaches our asymptotes. To the right of $x = 1$ there is no intercept, so we know the graph doesn't cross the x axis and therefore would have to be above the line $y = 2$.

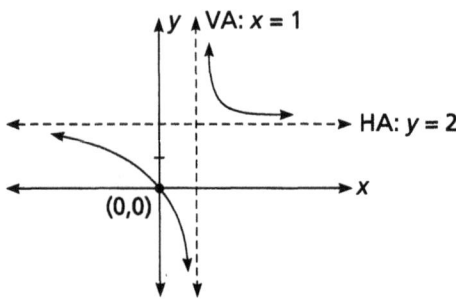

$$f(x) = \frac{2x}{x-1}$$

3. V.A.: $x = 2$ H.A.: $y = -1$
   x-intercept: $(-2, 0)$ y-intercept: $(0, 1)$

$2 + x = 0$ $f(0) = \dfrac{2+0}{2-0} = 1$

$x = -2$

Let's see what we have so far.

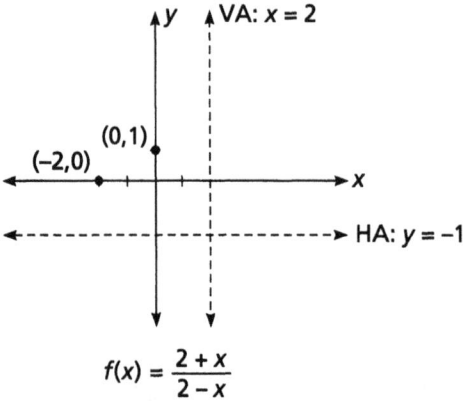

$$f(x) = \frac{2+x}{2-x}$$

We can see that the graph would have to be above $y = -1$ to the left of $x = 2$ because it has to pass through the intercepts. To the right of $x = 2$ it would have to be below $y = -1$ because there aren't any x-intercepts in that section of the graph.

Graphs of Functions   95

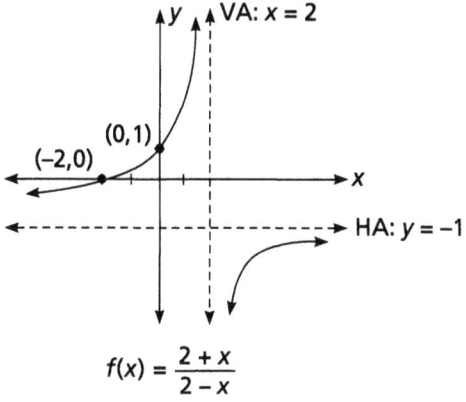

$$f(x) = \frac{2+x}{2-x}$$

4. V.A.: $x = -3$, $x = 3$ H.A.: $y = 0$

x-intercept: $(1,0)$ y-intercept: $\left(0, \frac{1}{9}\right)$

$x^2 - 9 = 0$ $f(0) = \frac{0-1}{0^2 - 9} = \frac{1}{9}$

$(x+3)(x-3) = 0$
$x = -3$, $x = 3$

Let's see how this looks so far.

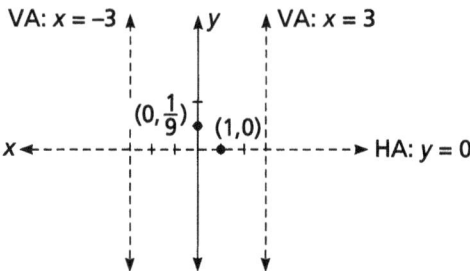

If we pick an x value to the left of −3 and substitute it into the function, we get a corresponding y value less than $y = 0$, so we know that part of the graph approaches the asymptotes below $y = 0$. If we pick an x value to the right of 3 and substitute it into the function, we get a corresponding y value greater than $y = 0$, so we know that part of the graph approaches the asymptotes above $y = 0$. Now we need to figure out what the y values are when the x values are between −3 and 3. Between −3 and 0 they're above $y = 0$, and between 0 and 3 they're below $y = 0$.

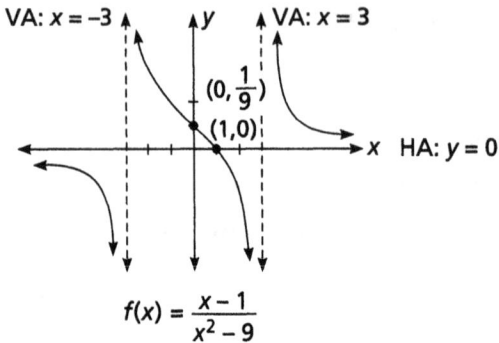

5. V.A.: none H.A.: $y = 1$
x-intercept: (0,0) y-intercept: (0,0)

$$f(0) = \frac{0^2}{0^2 + 9} = 0$$

All the y values will be positive because any number squared is positive and any positive number plus 9 will be positive.

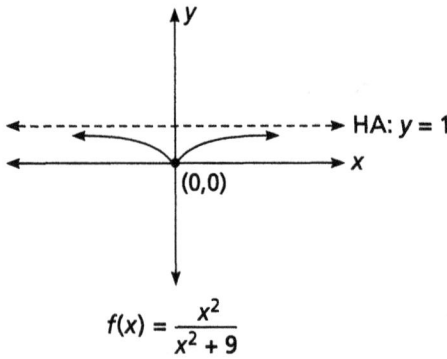

6. V.A.: $x = 0$, $x = 4$ H.A.: $y = 0$
x-intercept: (−1,0) y-intercept: none
$4(x + 1) = 0$
$x = -1$

Let's look at what we have so far. If we substitute an x value to the right of $x = 4$ into the function, we get a y value that's greater than $y = 0$. This part of the graph is above the horizontal asymptote. If we substitute an x value to the left of $x = 0$, the y value will be less than 0 until it approaches −1, where it crosses the x axis. This part of the graph is below $y = 0$ until −1; then it crosses the horizontal asymptote and approaches ∞. Between $x = 0$ and $x = 4$ any x value we substitute into the function will give us a corresponding y value that's negative. Keep in mind that the graph approaches its asymptotes.

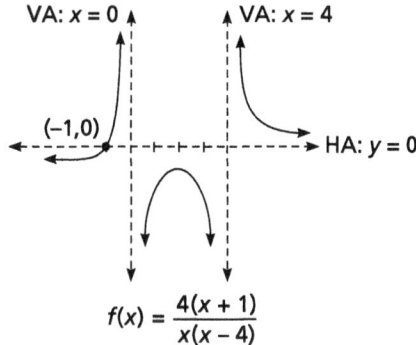

7. We can assume that any x value we substitute into the function will have a negative y value. This is the same problem as the first problem in this self-test except that it has a negative. Let's see how the negative will change the graph.

V.A.: $x = 0$ H.A.: $y = 0$
x-intercept: none y-intercept: none

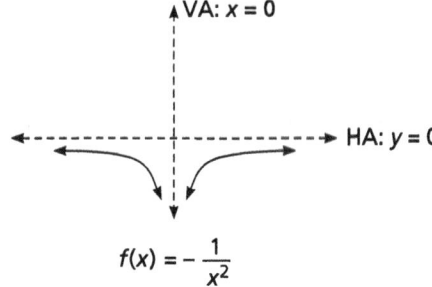

8. V.A.: $x = 2, x = -1$ H.A.: $y = 0$
x-intercept: (0,0) y-intercept: (0,0)

Any x value to the left of $x = -1$ will give a corresponding y value that's negative. Any x value to the right of $x = 2$ will give a corresponding y value that's positive. Between $x = -1$ and $x = 2$ the graph will cross the x axis at (0,0). Any x value between −1 and 0 will have a positive corresponding y value. Any x value between 0 and 2 will have a negative corresponding y value.

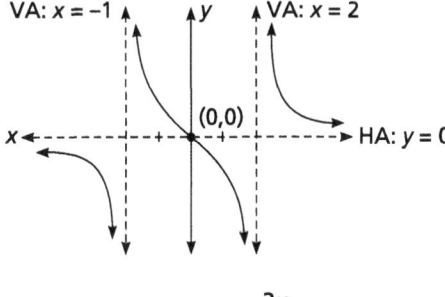

So far, none of the graphs we've sketched in this section had a numerator of higher degree than its denominator. This type of fraction is called an *improper fraction*; for example, $\frac{5}{2}$. When the degree of the numerator is greater than the degree of the denominator, we have a special type of asymptote called an *oblique asymptote*. Oblique asymptotes are not vertical or horizontal lines. In many cases they're not even straight lines. To find an oblique asymptote we divide the numerator by the denominator. Our quotient without its remainder is our oblique asymptote.

Example 29:

Find the oblique asymptote of the following function and sketch the graph:

$$f(x) = \frac{x^2 - x}{x + 1}$$

V.A.: $x = -1$ H.A.: none

x-intercepts: $(0,0), (1,0)$ y-intercept: $(0,0)$

To find the oblique asymptote of this improper fraction, we'll divide out the fraction.

$$\begin{array}{r} x - 2 \\ x+1 \overline{) x^2 - x } \\ \underline{x^2 + x} \\ -2x \\ \underline{-2x - 2} \\ 2 \end{array}$$

The oblique asymptote is the quotient. O.A.: $y = x - 2$. We can ignore the remainder $\frac{2}{x + 1}$ because as x gets infinitely larger, the remainder approaches 0. In this problem our O.A. is the straight line $y = x - 2$.

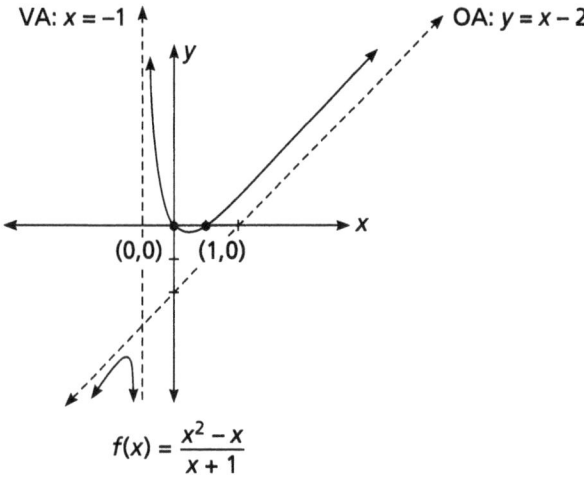

$$f(x) = \frac{x^2 - x}{x + 1}$$

Example 30:

Find the oblique asymptote of this improper fraction and sketch the graph:

$$f(x) = \frac{x^4 + 1}{x^2}$$

V.A.: $x = 0$
    x-intercept: none

H.A.: none
y-intercept: none

O.A.: $y = x^2$

This oblique asymptote is a parabola.

$$\begin{array}{r} x^2 \\ x^2 \overline{)x^4 + 1} \\ \underline{x^4} \\ 1 \end{array}$$

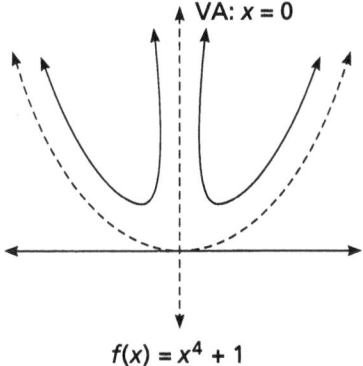

$$f(x) = x^4 + 1$$

SELF-TEST 10:

Graph the following functions, labeling asymptotes and intercepts:

1. $f(x) = \dfrac{x^2 - 3}{x}$
2. $f(x) = \dfrac{x^2 - x - 12}{x + 1}$
3. $f(x) = \dfrac{2x^5 - x^3 + 2}{x^3 - 1}$
4. $f(x) = \dfrac{x^4 - 1}{x^2 - 4}$

ANSWERS:

1. V.A.: $x = 0$
O.A.: $y = x$
x-intercept: $(-\sqrt{3}, 0), (\sqrt{3}, 0)$
H.A.: none
y-intercept: none

$x^2 - 3 = 0$
$x^2 = 3$
$x = \pm\sqrt{3}$

$f(0) = \dfrac{0^2 - 3}{0}$ is undefined.

$$\begin{array}{r} x \\ x\overline{)x^2 - 3} \\ \underline{x^2} \\ -3 \end{array}$$

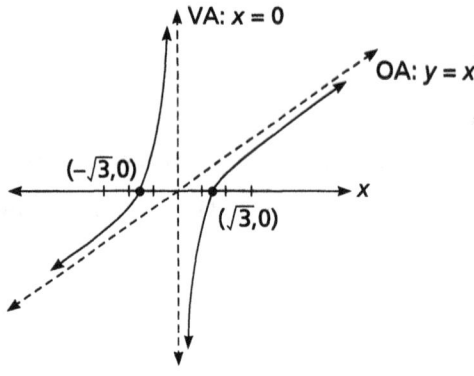

$f(x) = \dfrac{x^2 - 3}{x}$

2. V.A.: $x = -1$
O.A.: $y = x - 2$
x-intercept: $(4, 0), (-3, 0)$
$x^2 - x - 12 = 0$
$(x - 4)(x + 3) = 0$
$x = 4, x = -3$
H.A.: none
y-intercept: $(0, -12)$
$f(0) = -12$

$$\begin{array}{r} x - 2 \\ x + 1\overline{)x^2 - x - 12} \\ \underline{x^2 + x} \\ -2x - 12 \\ \underline{-2x - 2} \\ -10 \end{array}$$

Graphs of Functions 101

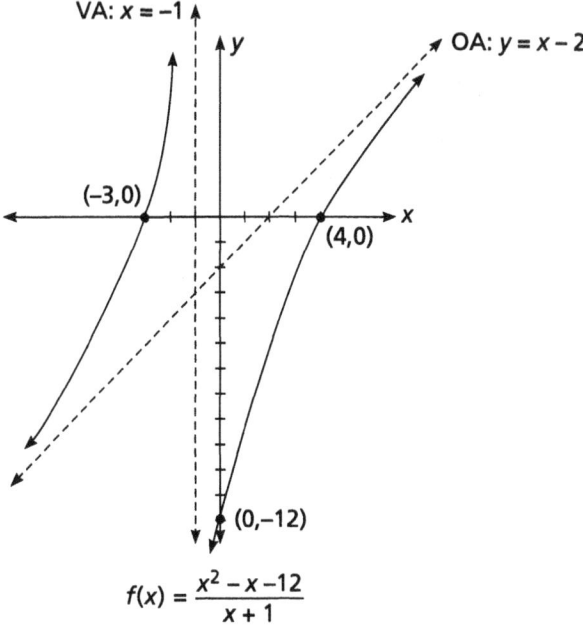

$$f(x) = \frac{x^2 - x - 12}{x + 1}$$

3. V.A.: $x = 1$ H.A.: none
O.A.: $y = 2x^2 - 1$ x-intercepts: $\approx(-1.11, 0)$ y-intercept: $(0,-2)$

$$\begin{array}{r} 2x^2 - 1 \\ x^3 - 1 \overline{\smash{)}2x^5 - x^3 + 2} \\ \underline{2x^5 - 2x^2} \\ -x^3 + 2x^2 + 2 \\ \underline{-x^3 + 1} \\ 2x^2 + 1 \end{array}$$

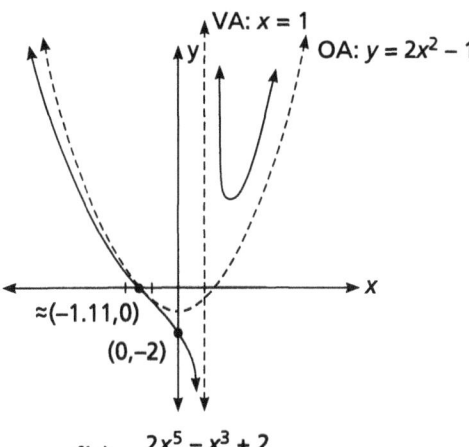

$$f(x) = \frac{2x^5 - x^3 + 2}{x^3 - 1}$$

4. V.A.: $x = 2, x = -2$ H.A.: none
O.A.: $y = x^2 + 4$ x-intercepts: $(1,0), (-1,0)$ y-intercept: $\left(0, \dfrac{1}{4}\right)$

$$\begin{array}{r} x^2 + 4 \\ x^2 - 4 \overline{\smash{)}x^4 - 1} \\ \underline{x^4 - 4x^2 } \\ 4x^2 - 1 \\ \underline{4x^2 - 16} \\ 15 \end{array}$$

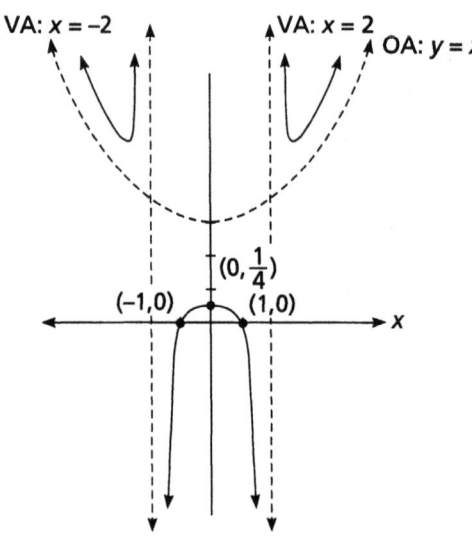

$$f(x) = \dfrac{x^4 - 1}{x^2 - 4}$$

4 Exponential and Logarithmic Functions

In this chapter we'll study two nonalgebraic functions: the exponential and the logarithmic. In our daily lives we encounter many examples of nonalgebraic functions. Some examples of exponential and logarithmic functions are continuously compounded interest, radioactive decay or growth, and carbon dating. In section 5 of this chapter we'll see examples of real-life applications of exponential and logarithm functions. By the time we've finished this chapter you'll be able to

- identify a function as exponential or logarithmic
- solve exponential and logarithmic functions
- solve word problems involving exponential and logarithmic functions

1 Exponential Functions

Before we give you a formal definition of an exponential equation, or function, let's look at two functions. The first is a standard algebraic function, $y = 2x$; the second is an exponential function, $y = 2^x$. In the first function, $y = 2x$, the variable x is used as a factor, not an exponent. In the second function, $y = 2^x$, the variable x is used as an exponent, not a factor. When the variable we're solving for is an exponent, the equation is called an exponential equation. So we know that $y = 2^x$ is an exponential function. Table 4.1 lists some values for each function. If you compare the values for $y = 2x$ and $y = 2^x$ in the table, you should notice that the values for 2^x increase at a much faster rate than for $2x$. This is called an exponential increase or growth.

Table 4.1

x	−2	−1	0	1	2	3	4	5	6	7	8
$2x$	−4	−2	0	2	4	6	8	10	12	14	16
2^x	$\frac{1}{4}$	$\frac{1}{2}$	1	2	4	8	16	32	64	128	256

The figures below allow us to compare the graphs of the functions.

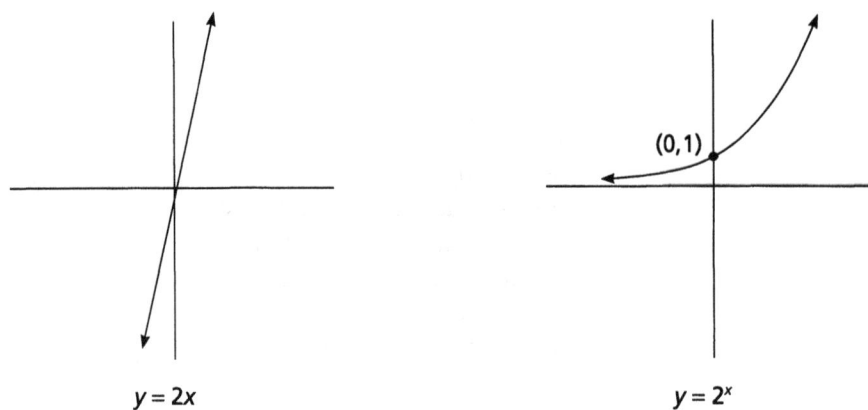

$y = 2x$ $y = 2^x$

An exponential function is a function of the form $f(x) = b^x$ where $b > 0$, $b \ne 1$, and x is any real number.

Let's see what happens to the graph of $y = 2^x$ if we change the x to $-x$. Table 4.2 lists some values for the function $y = 2^{-x}$. $y = 2^{-x}$ is the same as $y = \frac{1}{2^x}$.

Table 4.2

x	−4	−3	−2	−1	0	1	2	3	4
2^{-x}	16	8	4	2	1	$\frac{1}{2}$	$\frac{1}{4}$	$\frac{1}{8}$	$\frac{1}{16}$

The graph of $y = 2^{-x}$ is shown below. It's the same form as for $y = 2^x$ except that it's been rotated around the y-axis.

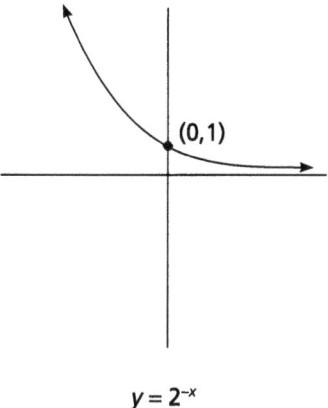

$y = 2^{-x}$

The figures below show the standard forms of the exponential functions $y = b^x$, $y = b^{-x}$, where $b > 0$. We'll use these graphs to state the domain, range, intercept, and asymptotes of the standard exponential functions.

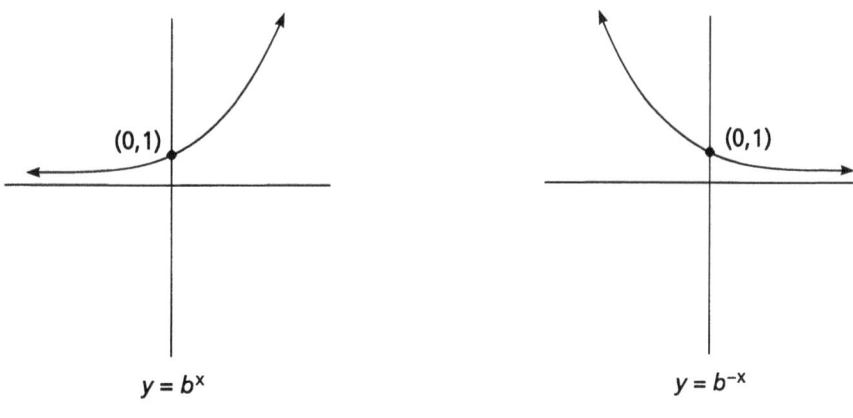

$y = b^x$

Domain: $(-\infty, \infty)$
Range: $(0, \infty)$
Intercept: $(0, 1)$
Increasing horizontal
Asymptote: $y = 0$

$y = b^{-x}$

Domain: $(-\infty, \infty)$
Range: $(0, \infty)$
Intercept: $(0, 1)$
Decreasing horizontal
Asymptote: $y = 0$

Now that you know what an exponential function is, it's time to learn how to solve an exponential equation. In this section we'll solve exponential equations that can be written with the same base. Once you've gotten the hang of these exponential equations you may find that they're actually fun to solve.

Example 1:
Solve $2^x = 8$.

The value for x is 3 because $2^3 = 8$.

Example 2:
Solve $5^x = 25$.

The value for x is 2 because $5^2 = 25$.

The solutions to examples 1 and 2 were pretty easy. Let's try a few problems where the solutions are a bit harder to find. When solutions are not obvious we usually have to rewrite the terms on opposite sides of the equal sign, so they have the same base. If the bases are equal, we can assume that the exponents are equal.

Example 3:
Solve $9^x = 81^x$.

As you can see, the solution to this equation is not as obvious as in the previous two examples. But don't worry; with a little work we'll make it obvious. The first thing we should do is rewrite both 9 and 81 with a common base. Nine can be written as 3^2, and 81 can be written as 3^4.

$9^x = 81^x$	
$(3^2)^x = (3^4)^x$	We replace 9 by 3^2 and 81 by 3^4.
$3^{2x} = 3^{4x}$	Multiply the exponents.
$2x = 4x$	If the bases are equal we can assume the exponents are equal.
$x = 0$	Solve for x.

Let's check $x = 0$ by substituting 0 in for x in the original equation.

$9^0 = 81^0$
$1 = 1$ $x = 0$ checks in the equation.

Let's try solving a few more exponential equations. Then it's your turn.

Exponential and Logarithmic Functions 107

Example 4:
Solve $4^{2x-7} = 64$.

Let's start by rewriting 64 as 4^3 so the same base is on opposite sides of the equal sign.

$4^{2x-7} = 4^3$
$2x - 7 = 3$ If the bases are the same, we can assume the exponents are equal.
$2x = 10$ Solve for x.
$x = 5$

Let's check the answer $x = 5$.

$4^{2(5)-7} = 64$
$4^3 = 64$

Example 5:
Solve $3^{x-1} = \dfrac{1}{27}$.

Twenty-seven is 3^3, so we'll rewrite this equation using 3 as the common base.

$3^{x-1} = \dfrac{1}{3^3}$
$3^{x-1} = 3^{-3}$ We bring the 3^3 in the denominator up to the numerator as 3^{-3}.
$x - 1 = -3$ The bases are the same, so we can assume the exponents are equal.
$x = -2$ Solve for x.

We'll leave the check for you.

Example 6:
Solve $5^{-\frac{t}{2}} = .2$.

This is the first problem with a decimal. When there's a decimal in the problem, it's usually best to re-write it as a fraction in lowest terms.

$5^{-\frac{t}{2}} = \dfrac{1}{5}$ We'll begin by writing $\dfrac{1}{5}$ as 5^{-1}.

$5^{-\frac{t}{2}} = 5^{-1}$ Now the bases are the same, so the exponents are equal

$-\dfrac{t}{2} = -1$ Solve for t.

$t = 2$

SELF-TEST 1:

Solve the following exponential equations:

1. $3^x = 243$
2. $8^x = 4$
3. $\left(\dfrac{3}{4}\right)^x = \dfrac{27}{64}$
4. $7^x = \dfrac{1}{49}$
5. $2^{3x-1} = 2^x$
6. $3^{5x+1} = 9^{2x}$
7. $25^{\sqrt{x}} = 5^x$
8. $\left(\dfrac{1}{16}\right)^{x-3} = 8^{2x-1}$
9. $\dfrac{4^x}{4^{2x}} = 64$
10. $\left(\dfrac{1}{2}\right)^x = 32$

ANSWERS:

1. $3^x = 243$
    $3^x = 3^5$
    $x = 5$

2. $8^x = 4$
    $(2^3)^x = 2^2$
    $2^{3x} = 2^2$
    $3x = 2$
    $x = \dfrac{2}{3}$

3. $\left(\dfrac{3}{4}\right)^x = \dfrac{27}{64}$
    $\left(\dfrac{3}{4}\right)^x = \left(\dfrac{3}{4}\right)^3$
    $x = 3$

4. $7^x = \dfrac{1}{49}$
    $7^x = \dfrac{1}{7^2}$
    $7^x = 7^{-2}$
    $x = -2$

5. $2^{3x-1} = 2^x$
    $3x - 1 = x$
    $2x = 1$
    $x = \dfrac{1}{2}$

6. $3^{5x+1} = 9^{2x}$
    $3^{5x+1} = (3^2)^{2x}$
    $3^{5x+1} = 3^{4x}$
    $5x + 1 = 4x$
    $x = -1$

7. $25^{\sqrt{x}} = 5^x$
    $(5^2)^{\sqrt{x}} = 5^x$
    $5^{2\sqrt{x}} = 5^x$
    $(2\sqrt{x})^2 = (x)^2$
    $4x = x^2$
    $0 = x^2 - 4x$
    $0 = x(x-4)$
    $x = 0, x = 4$

8. $\left(\dfrac{1}{16}\right)^{x-3} = 8^{2x-1}$
    $\left(\dfrac{1}{2^4}\right)^{x-3} = (2^3)^{2x-1}$
    $(2^{-4})^{x-3} = (2^3)^{2x-1}$
    $2^{-4x+12} = 2^{6x-3}$
    $-4x + 12 = 6x - 3$
    $15 = 10x$
    $\dfrac{15}{10} = x$
    $x = \dfrac{3}{2}$

9. $\dfrac{4^x}{4^{2x}} = 64$
    $4^{-x} = 4^3$
    $-x = 3$
    $x = -3$

10. $\left(\dfrac{1}{2}\right)^x = 32$
    $(2^{-1})^x = 2^5$
    $2^{-x} = 2^5$
    $-x = 5$
    $x = -5$

2 Logarithmic Functions

In chapter 3 we studied functions and their inverses. The inverse of an exponential function is a logarithmic function. When we discussed inverse functions we saw that the domain of the function is the range of the inverse function, and the range of the function is the domain of the inverse function (see page 45). To find the inverse of a function we interchanged the x's and the y's in the function and solved for x. We'll apply the same idea to the exponential function $y = 2^x$. If we interchange the x and the y we get $x = 2^y$, and y is called the logarithm of x with base 2, written $y = \log_2 x$. The logarithmic equation is read "y equals the log of x base 2." Log is the abbreviation for logarithm. If a base is not given, we assume it's base 10. Base 10 is referred to as the common log.

A logarithmic function is a function of the form $y = \log_b x$, where $b > 0$, $b \neq 1$; $y = \log_b x$ is equivalent to $x = b^y$.

Here is the graph of $y = \log_2 x$, which can be written as $x = 2^y$.

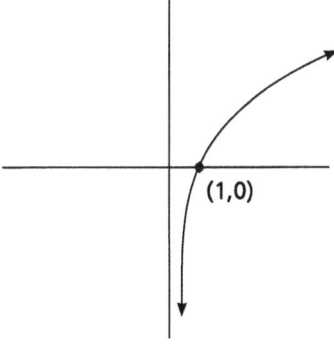

$y = \log_2 x$

Before we solve log equations it's important to be able to convert from log form to exponential form and from exponential form to log form. Take, for example, $x = 2^y$, which, as we've just learned, is equal to $y = \log_2 x$. When converting from exponential form to log form, the base of the exponential form (in this case, 2) becomes the base in the log form ($\log_2$). The exponent in the exponential form (y) becomes what the log is equal to ($y =$) in log form. The number the exponential form is equal to (x) becomes the number you take the log of ($\log_2 x$) in log form. It sounds confusing, but the next few examples should make it clearer.

PRECALCULUS

Example 7:
Convert the following exponential functions to log form.

Exponential Form	Logarithmic Form	
$2^3 = 8$	$3 = \log_2 8$	The bases stay the same in either form.
$\dfrac{1}{100} = 10^{-2}$	$\log_{10} \dfrac{1}{100} = -2$	The exponent becomes what the log is equal to.
$3^2 = 9$	$\log_3 9 = 2$	
$10^2 = 100$	$\log 100 = 2$	When a base in a log equation isn't written, it's assumed to be base 10.

Example 8:
Convert the following logarithmic functions to exponential form.

Logarithmic Form	Exponential Form
$\log \dfrac{1}{100} = -2$	$10^{-2} = \dfrac{1}{100}$
$\log_4 64 = 3$	$4^3 = 64$
$\log_{\frac{1}{4}} 16 = -2$	$\left(\dfrac{1}{4}\right)^{-2} = 16$
$\log_{.1} \dfrac{1}{100} = 2$	$.1^2 = \dfrac{1}{100}$

SELF-TEST 2:

Write the exponential equation in logarithmic form:

1. $2^5 = 32$
2. $10^3 = 1000$
3. $9^0 = 1$
4. $5^{-2} = \dfrac{1}{25}$
5. $\left(\dfrac{1}{3}\right)^3 = \dfrac{1}{27}$

ANSWERS:

1. $\log_2 32 = 5$
2. $\log 1000 = 3$
3. $\log_9 1 = 0$
4. $\log_5 \dfrac{1}{25} = -2$
5. $\log_{\frac{1}{3}} \dfrac{1}{27} = 3$

Exponential and Logarithmic Functions 111

SELF-TEST 3: Write the logarithmic equation in exponential form.

1. $\log_2 64 = 6$
2. $\log_8 8 = 1$
3. $\log .01 = -2$
4. $\log_8 1 = 0$
5. $\log_{\frac{1}{3}}\left(\frac{1}{9}\right) = 2$

ANSWERS:
1. $2^6 = 64$
2. $8^1 = 8$
3. $10^{-2} = 0.01$
4. $8^0 = 1$
5. $\left(\frac{1}{3}\right)^2 = \left(\frac{1}{9}\right)$

3 Properties of Logarithmic Functions

For us to solve some log equations we sometimes have to use properties of log functions to rewrite them in a form that makes it easier. We will usually refer to this as expanding, or writing as a single log. In the next section we'll concentrate on solving log equations using three basic properties of logs.

Property 1: The Multiplication Property
For any positive real numbers x, y, and b, $b \neq 1$, $\log_b xy \Leftrightarrow \log_b x + \log_b y$.

If we wanted to state this property in English, we would say the log of a product is equal to the sum of the logs. Let's look at some examples of this property.

Example 9:
$\log_2 3z = \log_2 3 + \log_2 z$

Example 10:
$\log st = \log s + \log t$

When we stated the multiplication property, notice we used the symbol $\Leftrightarrow$. This symbol means you can write what's on the left as what's on the right, or you can write what's on the right as what's on the left. That means we could apply the multiplication property in reverse as $\log_b x + \log_b y = \log_b xy$. In words, we would say the sum of logs is equal to the log of the product. Let's rewrite examples 9 and 10 using this form of the multiplication property.

Example 9:

$\log_2 3 + \log_2 z = \log_2 3z$

Example 10:

$\log s + \log t = \log st$

Now that you know how to apply the multiplication property, it's time to learn the division property.

Property 2: The Division Property

For any positive real numbers, x, y, and b, b ≠ 1, $\log_b \frac{x}{y} \Leftrightarrow \log_b x - \log_b y$.

If we wanted to state this property in English, we would say the log of a quotient is equal to the difference of the logs. As in the case of the multiplication property, we used the ⇔ symbol because the division property also can be applied in the reverse order, $\log_b x - \log_b y \Leftrightarrow \log_b \frac{x}{y}$. We could say the difference of the logs is equal to the log of the quotient. Watch out for a very common error, $\log_b \frac{x}{y} \neq \frac{\log_b x}{\log_b y}$. Let's look at a couple of examples of how to apply this property.

Example 11:

$\log_3 \frac{6}{x} = \log_3 6 - \log_3 x$

Example 12:

$\log_5 w - \log_5 4 = \log_5 \frac{w}{4}$

Now that you know how to apply the division property, it's time to learn the power property.

Property 3: The Power Property

For any positive real numbers x and b, b ≠ 1, and for any real number r, $\log_b x^r \Leftrightarrow r \log_b x$.

If we wanted to state this property in English, we would say the log of a term to a power becomes the power times the log of the term. The exponent becomes a coefficient. For this property we also used the ⇔ symbol, which tells us that this property also can be reversed: $r \log_b x \Leftrightarrow \log_b x^r$. Let's look at a couple of examples.

Example 13:

$\log_2 c^3 = 3 \log_2 c$ The exponent becomes a coefficient.

Example 14:

$6 \log t = \log t^6$ The coefficient becomes an exponent.

Now it's time for you to try a few problems.

SELF-TEST 4: Apply one of the log properties to the following problems:

1. $\log_4 5x + \log_4 6x$
2. $\log_2 4x - \log_2 16y$
3. $2\log y$
4. $\log_7 c^4$
5. $\log_3 rs$
6. $\log \dfrac{4t}{3s}$

ANSWERS:

1. $\log_4 30x^2$
2. $\log_2 \dfrac{4x}{16y} = \log_2 \dfrac{x}{4y}$
3. $\log y^2$
4. $4 \log_7 c$
5. $\log_3 r + \log_3 s$
6. $\log 4t - \log 3s$

Generally it's necessary to apply more than one of the properties of logs to a problem. If a log has a coefficient, you must apply the power property before you apply the multiplication or division property. In the next example we'll apply the power property first, then the multiplication property.

Example 15:
Expand $6 \log_{16} 2x$.

$6 \log_{16} 2x = \log_{16} (2x)^6 = \log_{16} 64x^6 = \log_{16} 64 + \log_{16} x^6$

In the next example we'll apply the power property first, then the division property. This time, instead of expanding, we'll condense, or we could say write as a single log.

Example 16:
Write as a single log with a coefficient of 1: $4 \log s - 2 \log t$.

$4 \log s - 2 \log t = \log s^4 - \log t^2 = \log \dfrac{s^4}{t^2}$

Example 17:
Expand $\log_8 \sqrt{a^2 b}$.

$\log_8 \sqrt{a^2 b} = \log_8 (a^2 b)^{\frac{1}{2}} = \log_8 ab^{\frac{1}{2}} = \log_8 a + \log_8 b^{\frac{1}{2}} = \log_8 a + \dfrac{1}{2} \log_8 b$

Example 18:

Write as a single log with a coefficient of 1: $\frac{1}{4}(\log x - 5\log y + \log z)$.

$$\frac{1}{4}(\log x - \log y^5 + \log z) = \frac{1}{4}\left(\log \frac{x}{y^5} + \log z\right) = \frac{1}{4}\log \frac{xz}{y^5} = \log \sqrt[4]{\frac{xz}{y^5}}$$

SELF-TEST 5:

Write the following as a single log:

1. $2\log_3 x + \log_3 y$
2. $3(\log_2 x + 2\log_2 y - \log_2 z)$
3. $2\log x - 3(\log y - \log z)$
4. $\frac{1}{2}\log_b x - \frac{2}{3}\log_b y + \frac{1}{2}\log_b z$

Expand the following:

5. $\log_9\left(\frac{a}{b^2}\right)$
6. $\log \sqrt{x^4 y}$
7. $\log_3 \sqrt[4]{\frac{s^2 t}{u}}$
8. $\log_7 (x^2 y)^3$

ANSWERS:

1. $2\log_3 x + \log_3 y = \log_3 x^2 + \log_3 y = \log_3 x^2 y$

2. $3(\log_2 x + 2\log_2 y - \log_2 z) = 3(\log_2 x + \log_2 y^2 - \log z) = 3(\log_2 xy^2 - \log z)$
$= 3\left(\log_2 \frac{xy^2}{z}\right) = \log_2 \frac{x^3 y^6}{z^3}$

3. $2\log x - 3(\log y - \log z) = \log x^2 - 3\left(\log \frac{y}{z}\right) = \log x^2 - \log \frac{y^3}{z^3} = \log \frac{x^2 z^3}{y^3}$

4. $\frac{1}{2}\log_b x - \frac{2}{3}\log_b y + \frac{1}{2}\log_b z = \log_b x^{\frac{1}{2}} - \log_b y^{\frac{2}{3}} + \log_b z^{\frac{1}{2}} = \log_b \frac{x^{\frac{1}{2}}}{y^{\frac{2}{3}}} + \log_b z^{\frac{1}{2}} = \log_b \frac{\sqrt{xz}}{\sqrt[3]{y^2}}$

5. $\log_9\left(\frac{a}{b^2}\right) = \log_9 a - \log_9 b^2 = \log_9 a - 2\log_9 b$

6. $\log \sqrt{x^4 y} = \log (x^4 y)^{\frac{1}{2}} = \frac{1}{2}\log (x^4 y) = \frac{1}{2}[\log x^4 + \log y] = \frac{1}{2}[4\log x + \log y] = 2\log x + \frac{1}{2}\log y$

7. $\log_3 \sqrt[4]{\frac{s^2 t}{u}} = \log_3 \left(\frac{s^2 t}{u}\right)^{\frac{1}{4}} = \frac{1}{4}\log_3 \left(\frac{s^2 t}{u}\right) = \frac{1}{4}[\log_3 s^2 t - \log_3 u] = \frac{1}{4}[2\log_3 s + \log_3 t - \log_3 u]$

8. $\log_7 (x^2 y)^3 = 3\log_7 x^2 y = 3\log_7 x^2 + 3\log_7 y = 6\log_7 x + 3\log_7 y$

4 Solving Logarithmic Equations

Now that you know how to apply the properties of logs, we're almost ready to solve log equations. There's just one more thing you should be aware of. There's a special log called the natural log, written ln. What makes ln special is that its base is assumed to be base e; e is called Euler's constant. Somewhere on your calculator will be an e button and an ln button. For us to clearly explain Euler's constant, we would have to use some techniques of calculus, so you'll have to wait to read about this later. For now, let's just say the use of the natural log will help us solve log equations and will be necessary in the applications section of this chapter. The following is a step-by-step procedure of how to solve a log equation.

Steps to solve a log equation:
1. Move all logs to one side of the equation.
2. Write as a single log.
3. Convert from log form to exponential form.
4. Solve the remaining equation.
5. Check to make sure your answers are in the domain of the function (you can only take the log of a positive number).

Example 19:
Solve $\log x + \log(x + 3) = 1$.

Solution:
All the logs are already on one side, so our first step will be to write this equation as a single log.

$\log x + \log(x + 3) = 1$
$\quad \log x(x + 3) = 1$ Convert from log form to exponential form.
$\quad\quad\quad 10^1 = x^2 + 3x$ If no base is written, it's base 10. Solve the remaining equation.
$\quad\quad x^2 + 3x - 10 = 0$ Factor and solve for x.
$\quad\quad (x - 2)(x + 5) = 0$
$x = 2, x = -5$ When you substitute -5 into the function you would take the log of a negative number. This is not acceptable, so $x = 2$ is the only acceptable answer.
$x = 2$

Example 20:

Solve $\log_4 x = \log_4 2 + \log_4 3$.

Solution:

$\log_4 x - \log_4 2 - \log_4 3 = 0$ Our first step is to move all logs to one side.

$\log_4 \frac{x}{2} - \log_4 3 = 0$ We used the properties of logs to condense to a single log.

$\log_4 \frac{x}{6} = 0$

$4^0 = \frac{x}{6}$ We converted from log form to exponential form.

$1 = \frac{x}{6}$

$x = 6$ We solved for x.

SELF-TEST 6:

Solve the following equations:

1. $\log_5 (x + 2) = 1$

2. $\log_3 x = \log_3 2 + \log_3 (x^2 - 3)$

3. $\frac{1}{2} \log (x + 2) + \log 5 = 1$

4. $\ln 5 - \ln x = -1$

5. $3 \ln 2 + \ln (x - 1) = \ln 24$

ANSWERS:

1. $\log_5 (x + 2) = 1$
$5^1 = x + 2$
$x = 3$

2. $\log_3 x = \log_3 2 + \log_3 (x^2 - 3)$
$\log_3 x - \log_3 2 - \log_3 (x^2 - 3) = 0$
$\log_3 \frac{x}{2(x^2 - 3)} = 0$
$3^0 = \frac{x}{2(x^2 - 3)}$
$1 = \frac{x}{2(x^2 - 3)}$
$2(x^2 - 3) = x$
$2x^2 - x - 6 = 0$
$(2x + 3)(x - 2) = 0$
$x = -\frac{3}{2}, x = 2$

If we substitute $x = -\frac{3}{2}$ into the original function it won't check, so the only answer is $x = 2$.

3. $\frac{1}{2}\log(x+2) + \log 5 = 1$

$\log(x+2)^{\frac{1}{2}} + \log 5 = 1$

$\log 5\sqrt{x+2} = 1$

$10^1 = 5\sqrt{x+2}$

$2 = \sqrt{x+2}$

$4 = x+2$

$2 = x$

4. $\ln 5 - \ln x = -1$

$\ln \frac{5}{x} = -1$

$e^{-1} = \frac{5}{x}$

$\frac{1}{e} = \frac{5}{x}$

$x = 5e$

$x \approx 13.5914$

5. $3\ln 2 + \ln(x-1) = \ln 24$

$3\ln 2 + \ln(x-1) - \ln 24 = 0$

$\ln 2^3 + \ln(x-1) - \ln 24 = 0$

$\ln 8(x-1) - \ln 24 = 0$

$\ln \frac{8(x-1)}{24} = 0$

$\ln \frac{x-1}{3} = 0$

$e^0 = \frac{x-1}{3}$

$1 = \frac{x-1}{3}$

$x - 1 = 3$

$x = 4$

All the exponential equations we solved in section 1 of this chapter were easy to rewrite with a common base and solve. In the following section on applications, we'll encounter equations where we can't rewrite the terms with the same base—for example, $2^x = 3$. We can't rewrite 2 as a base of 3, or 3 as a base of 2. When this occurs we'll take the natural log of both sides of the equation, $\ln 2^x = \ln 3$. Now we'll solve for x.

Example 21:

Solve $2^x = 3$.

$2^x = 3$

$\ln 2^x = \ln 3$

$x \ln 2 = \ln 3$ To solve for x we move the x to a coefficient position.

$x = \frac{\ln 3}{\ln 2}$

$x \approx 1.585$

Example 22:

Solve $4^{2x} = 7$.

$4^{2x} = 7$

$\ln 4^{2x} = \ln 7$

$2x \ln 4 = \ln 7$ We'll start by moving the exponent to a coefficient position.

$2x = \dfrac{\ln 7}{\ln 4}$

$x = \dfrac{\ln 7}{2 \ln 4}$

$x \approx .7018$

Example 23:

Solve $e^{3x} = 21$.

$e^{3x} = 21$

$\ln e^{3x} = \ln 21$ Start by taking the ln of both sides of the equation.

$3x = \ln 21$ Now we're going to use a shortcut. When the base and the term you're taking the log of have the same base—in this case, base *e*—it simplifies down to just the exponent of the term; for example, $\log_2 2^3 = 3$, $\log 10^4 = 4$, $\log_4 64 = \log_4 4^3 = 3$, $\ln e^6 = 6$, $\log_5 25 = 5^2 = 2$.

$x = \dfrac{\ln 21}{3}$

$x \approx 1.0148$

We'll use this shortcut repeatedly in the next section. Now it's time for you to try a few problems.

SELF-TEST 7:

Solve the following equations:

1. $10^x = 20$
2. $5^{-\frac{x}{3}} = 29$
3. $2^{3-x} = 50$
4. $3e^{4x} = 9$
5. $10(100 - e^{\frac{x}{2}}) = 40$

ANSWERS:

1. $10^x = 20$
$\ln 10^x = \ln 20$ Take the ln of both sides of the equation.
$x \ln 10 = \ln 20$ Move the *x* to a coefficient position.
$x = \dfrac{\ln 20}{\ln 10}$ Divide both sides by ln 10.
$x \approx 1.3010$

Exponential and Logarithmic Functions

2. $5^{-\frac{y}{3}} = 29$

$\ln 5^{-\frac{y}{3}} = \ln 29$ — Take the ln of both sides of the equation.

$-\frac{y}{3} \ln 5 = \ln 29$ — Move $-\frac{y}{3}$ to a coefficient position.

$-\frac{y}{3} = \frac{\ln 29}{\ln 5}$ — Divide both sides by ln 5.

$y = -\frac{3 \ln 29}{\ln 5}$ — Multiply both sides by –3.

$y \approx 6.2767$

3. $2^{3-x} = 50$

$\ln 2^{3-x} = \ln 50$ — Take the ln of both sides of the equation.

$(3 - x) \ln 2 = \ln 50$ — Move the $(3 - x)$ to a coefficient position.

$3 - x = \frac{\ln 50}{\ln 2}$ — Divide both sides by ln 2.

$-x = \frac{\ln 50}{\ln 2} - 3$ — Move the 3 over to the other side as –3.

$x = -\frac{\ln 50}{\ln 2} + 3$ — Multiply both sides by –1.

$x \approx -2.6439$

4. $3e^{4x} = 9$

$e^{4x} = \frac{9}{3}$ — Divide both sides by 3.

$e^{4x} = 3$

$\ln e^{4x} = \ln 3$ — Take the ln of both sides.

$4x = \ln 3$ — Both bases are e, so use the shortcut from example 23.

$x = \frac{\ln 3}{4}$ — Divide both sides by 4.

$x \approx .2747$

5. $10(100 - e^{\frac{x}{2}}) = 40$

$100 - e^{\frac{x}{2}} = 4$ — Divide both sides by 10.

$-e^{\frac{x}{2}} = -96$ — Subtract 100 from both sides.

$e^{\frac{x}{2}} = 96$ — Multiply both sides by –1.

$\ln e^{\frac{x}{2}} = \ln 96$ — Take the ln of both sides.

$\frac{x}{2} = \ln 96$ — Both bases are e, so use the shortcut from example 23.

$x = 2 \ln 96$ — Multiply both sides by 2.

$x \approx 9.1287$

5 Applications

Now it's time for our favorite part of the chapter—applications, where we can see how what we've learned can be applied to the real world. The following examples will show us how to determine which investments are better than others. When solving problems involving money, we have two formulas to choose from. The first is used when the money is not compounded continuously.

1. $A = P\left(1 + \dfrac{r}{n}\right)^{nt}$ *A* is the amount present after *t* years.
P is the amount of dollars invested.
r is the interest rate (use decimal form).
n is the number of times the amount invested is compounded per year.
t is the number of years.

2. $A = Pe^{rt}$ *e* is Euler's constant.

The second is used when money is compounded continuously.

Example 24:

$50,000 is deposited into a bank account at 7 percent interest for ten years, compounded quarterly. How much money will there be in the account after ten years? How much interest was made in the ten years?

Solution:

To figure out this amount we have to use the formula $A = P(1 + \dfrac{r}{n})^{nt}$ because the example doesn't say compounded continuously; it's compounded quarterly. In this problem we want to find *A*. We'll substitute the following values into the equation: $P = \$50{,}000$, $r = .07$ (always substitute the interest rate in decimal form), $n = 4$ (because quarterly means four times a year), $t = 10$.

$$A = P\left(1 + \dfrac{r}{n}\right)^{nt} = 50{,}000\left(1 + \dfrac{.07}{4}\right)^{4(10)} = \$100{,}080\text{, the amount in the account after ten years; \$50{,}080 is the amount of interest.}$$

Let's see how much of a difference it would make if the money had been compounded continuously.

Example 25:

$50,000 is deposited into a bank account at 7 percent interest for ten years, compounded continuously. How much money will there be in the account after ten years? How much interest was made in the ten years? How much more interest was accrued because the investment was compounded continuously, not quarterly, as in example 24?

Solution:

This time we'll use the equation $A = Pe^{rt}$ because it's compounded continuously. The values for P, r, and t are as in example 24. $A = Pe^{rt} = 50{,}000e^{.07(10)} = \$100{,}688$, the amount in the account after ten years; $\$50{,}688$ is the amount of interest. If you compound continuously instead of quarterly, you would make $\$608$ more in interest.

In the previous two examples we were asked to find the amount in an account after a certain amount of time had passed. The following examples will ask us to solve for other information.

Example 26:

How long would it take for an investment to double if the amount invested is at 6 percent compounded daily?

Solution:

We aren't given P, but we do know it will be doubled, which would be $2P$. So we'll replace A by $2P$, r by .06, and n by 365. We always use 365 for the number of days in a year. We also know we have to use the first equation because this is not compounded continuously.

$$A = P\left(1 + \frac{r}{n}\right)^{nt}$$ We can replace A by $2P$ because the investment is doubled.

$$2P = P\left(1 + \frac{r}{n}\right)^{nt}$$

$$2 = \left(1 + \frac{.06}{365}\right)^{365t}$$ We can divide both sides of the equation by P.

$$\ln 2 = \ln\left(1 + \frac{.06}{365}\right)^{365t}$$ Whenever the variable we're solving for is in an exponential position, we'll take the natural log of both sides of the equation.

$$\ln 2 = 365t \ln\left(1 + \frac{.06}{365}\right)$$ Move the $365t$ to a coefficient position.

$$\frac{\ln 2}{365 \ln\left(1 + \frac{.06}{365}\right)} = t.$$ Divide both sides to solve for t.

$t \approx 11.5534$ years to double the investment.

Example 27:

At what interest rate would it take for an initial investment of $2,000 to triple in ten years if it's compounded continuously?

Solution:

We'll use the second equation, $A = Pe^{rt}$, because it's compounded continuously. $A = 6,000$ because the 2,000 is tripled, $P = 2,000$, and $t = 10$.

$A = Pe^{rt}$
$6,000 = 2,000e^{10r}$
$3 = e^{10r}$ — Divide both sides by 2,000.
$\ln 3 = \ln e^{10r}$ — Take the ln of both sides and use the shortcut we use when the bases are the same.
$\ln 3 = 10r$
$r = \dfrac{\ln 3}{10}$ — Divide both sides by 10.
$r \approx .10986$ — The interest rate would have to be about 11 percent for the investment to triple in ten years.

Now it's your turn to try a few.

SELF-TEST 8:

1. How much money would you owe the bank if you borrowed $175,000 for thirty years at 8 percent:

    a. compounded monthly?   b. compounded continuously?

2. How much money would you need to invest at a 9 percent interest rate compounded semiannually if you wanted to have $10,000 available in fifteen years?

3. You have to choose between two investment options for $10,000 of your money. The first is at an interest rate of 5 percent compounded continuously, the second is at an interest rate of 5.5 percent compounded

Exponential and Logarithmic Functions 123

quarterly, and both options are for twelve years. Which one is the better option, and how much more money would that option make for you?

4. How long will it take an amount of money invested at 7.2 percent compounded continuously to double?

ANSWERS:

1. $P = 175{,}000$, $t = 30$, $r = .08$, $A = ?$
 a. $n = 12$
 $$A = P\left(1 + \frac{r}{n}\right)^{nt}$$
 $$A = 175{,}000\left(1 + \frac{.08}{12}\right)^{12(30)}$$
 $A = \$1{,}913{,}753$

 b. $A = Pe^{rt}$
 $A = 175{,}000 e^{.08(30)}$
 $A = \$1{,}929{,}056$

2. $P = ?$, $r = .09$, $n = 2$, $A = 10{,}000$, $t = 15$
 $$A = P\left(1 + \frac{r}{n}\right)^{nt}$$
 $$10{,}000 = P\left(1 + \frac{.09}{2}\right)^{2(15)}$$
 $$\frac{10{,}000}{\left(1 + \frac{.09}{2}\right)^{30}} = P$$
 $P = \$2{,}670$

3. First option: $P = 10{,}000$, $r = .05$, $t = 12$
 $A = Pe^{rt}$
 $A = 10{,}000 e^{.05(12)}$
 $A = \$18{,}221.20$
 Second option: $r = .055$, $n = 4$
 $$A = P\left(1 + \frac{r}{n}\right)^{nt}$$
 $$A = 10{,}000\left(1 + \frac{.055}{4}\right)^{4(12)}$$
 $A = \$19{,}261.10$
 You will earn $1,039.92 more if you choose the second option.

4. $t = ?$, $P = P$, $A = 2P$, $r = .072$
 $A = Pe^{rt}$
 $2P = Pe^{.072t}$
 $2 = e^{.072t}$
 $\ln 2 = \ln e^{.072t}$
 $\ln 2 = .072t$
 $$\frac{\ln 2}{.072} = t$$
 $t \approx 9.627$

Now that you know how to figure out your best investments, let's look at some exponential growth and decay applications of log equations. The formula we'll use for this type of problem is:

$A = A_0 e^{kt}$ A is the size at time t.
 A_0 is the initial amount.
 t is the amount of time.
 k is called the constant of proportionality. It's the rate at which something grows or decays. If $k > 0$, we have a growth; if $k < 0$, we have a decay.

Example 28:

A colony of bacteria increases exponentially. If the number of bacteria doubles in three hours, how long will it take for the size of the colony to triple?

Solution:

We'll use the formula $A = A_0 e^{kt}$. We know that the initial amount doubles in three hours, so we know $A = 2A_0$, $t = 3$. We don't know the value of k. In most of these equations we'll have to begin by solving for k.

$$A = A_0 e^{kt}$$
$$2A_0 = A_0 e^{3k}$$
$$2 = e^{3k}$$
$$\ln 2 = \ln e^{3k}$$
$$\ln 2 = 3k$$
$$\frac{\ln 2}{3} = k$$
$$k \approx .231$$

Once we know the value of k, we can finish the problem.

$$A = A_0 e^{kt}$$
$$3A_0 = A_0 e^{.231t}$$
$$3 = e^{.231t}$$
$$\ln 3 = \ln e^{.231t}$$
$$\ln 3 = .231t$$
$$\frac{\ln 3}{.231} = t$$
$$t \approx 4.756$$

It will take approximately 4.756 hours for the bacteria to triple.

Some substances have what's called a half-life, which is the time it takes for half of a particular substance to decay. We use the knowledge of a substance's half-life to perform an operation called carbon dating to estimate the age of an object. In carbon dating we use the fact that all living organisms contain two kinds of carbon: carbon-12 (a stable carbon), and carbon-14 (a radioactive carbon, with a half-life of 5,600 years). While an organism is living, the ratio of carbon-12 to carbon-14 is constant. But when an organism dies, the original amount of carbon-12 present remains unchanged, whereas the amount of carbon-14 begins to decrease. This change in the amount of carbon-14 present relative to the amount of carbon-12 present makes it possible to calculate when an organism died.

Example 29:

Traces of burned wood found along with ancient stone tools in an archaeological dig in Chile were found to contain approximately 1.67 percent of the original amount of carbon-14. If the half-life of carbon-14 is 5,600 years, approximately when was the tree cut and burned?

Solution:

First we need to find the k (constant of proportionality). We'll let $A = \frac{1}{2} A_0$ because we're using a half-life of 5,600, $t = 5,600$.

$$A = A_0 e^{kt}$$
$$\frac{1}{2} A_0 = A_0 e^{5600k}$$
$$\frac{1}{2} = e^{5600k}$$
$$\ln .5 = 5,600k$$
$$\frac{\ln .5}{5,600} = k$$
$$k \approx -.000124$$

Now that we know k, we can finish the problem.

$$A = A_0 e^{-.000124t}$$
$$.0167 A_0 = A_0 e^{-.000124t}$$
$$.0167 = e^{-.000124t}$$
$$\ln .0167 = -.000124t$$
$$\frac{\ln .0167}{-.000124} = t$$
$$t \approx 33,002.8 \text{ years.}$$

We'll let $A = .0167 A_0$ because 1.67 percent of the original amount of carbon-14 is present.

SELF-TEST 9:

1. The population of a colony of mosquitoes grows at an exponential rate. If there are 2,000 mosquitoes initially, and 3,200 after two days, what is the size of the colony after three days? How long is it until there are 15,000 mosquitoes?

2. The population of a town decreases at an exponential rate. If it decreased from 500,000 to 450,000 from 1999 to 2001, what will the population be in 2005?

3. A fossilized leaf contains 70 percent of its normal amount of carbon-14. How old is the fossil?

4. The half-life of radium is 1,690 years. If 10 grams are present now, how much will be present in 50 years?

ANSWERS:

1. $A = A_0 e^{kt}$, $A_0 = 2{,}000$, $A = 3{,}200$, $t = 2$
$3{,}200 = 2{,}000 e^{2k}$
$\dfrac{3{,}200}{2{,}000} = e^{2k}$
$\ln \dfrac{8}{5} = 2k$

$k = \dfrac{\ln \dfrac{8}{5}}{2}$
$k \approx .235$

$A = ?$, $t = 3$
$A = 2{,}000 e^{.235(3)}$
$A \approx 4{,}047.69$
$t = ?$, $A = 15{,}000$
$15{,}000 = 2{,}000 e^{.235t}$
$\dfrac{15{,}000}{2{,}000} = e^{.235t}$
$\ln \dfrac{15}{2} = .235 t$
$t = \dfrac{\ln \dfrac{15}{2}}{.235} \approx 8.5741$

2. $A_0 = 500{,}000$ $A = 450{,}000$ $t = 2$
$A = A_0 e^{kt}$
$450{,}000 = 500{,}000 e^{2k}$
$\dfrac{450{,}000}{500{,}000} = e^{2k}$
$\ln \dfrac{9}{10} = 2k$

$k = \dfrac{\ln \dfrac{9}{10}}{2} \approx -.0527$

$A = ?$ $t = 6$
$A = 500{,}000 e^{-.0527(6)}$
$A \approx 364{,}457$

3. From example 29 we know that the half-life of carbon-14 is 5600 years and $k \approx -.000124$.
$.7 A_0 = A_0 e^{-.000124 t}$
$\ln .7 = -.000124 t$
$t = \dfrac{\ln .7}{-.000124} \approx 2876.41$

4. $\dfrac{1}{2} A_0 = A_0 e^{1690 k}$
$\dfrac{1}{2} = e^{1690 k}$
$\ln \dfrac{1}{2} = 1690 k$

$k = \dfrac{\ln \dfrac{1}{2}}{1690} \approx -.00041$
$A = 10 e^{-.00041(50)}$
$A \approx 9.7971$ grams

5 Trigonometry

We will use trigonometry to develop ways to measure sides and angles of triangles. This will help us to solve problems involving surveying, navigation, optics, the motion of objects, finding distances, and many other applications. An understanding of trigonometry and how to apply it is essential to the study of calculus. In this chapter you will learn:

- the basic angles
- the basic trigonometric functions and their inverses
- graphs of trigonometric functions
- how to solve right triangles
- how to use your knowledge of the trigonometric functions to solve application problems involving right triangles

1 Angles and Their Measure

We'll begin this section by drawing what's referred to as the standard unit circle, which is shown on page 128. The unit circle has a radius of 1 and is centered at the origin. A complete revolution around the circle is 360 degrees, which is written 360°. The unit circle is divided into four counterclockwise regions, called quadrants. They are labeled using the Roman numerals I, II, III, and IV.

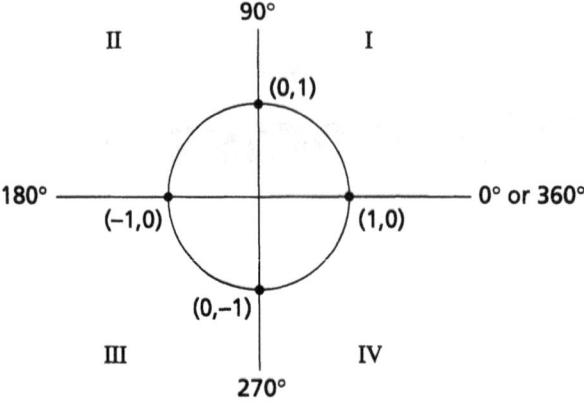

All angles have an initial side and a terminal side, as illustrated below. If the angle is formed by moving counterclockwise, the angle is positive; if it's formed by moving clockwise, the angle is negative. If two angles share the same terminal side, they are called coterminal angles. One complete revolution around a circle is 360°. Coterminal angles are formed by adding or subtracting 360 or a multiple of 360. In the figure below, 90°, −270°, and 450° are all coterminal angles.

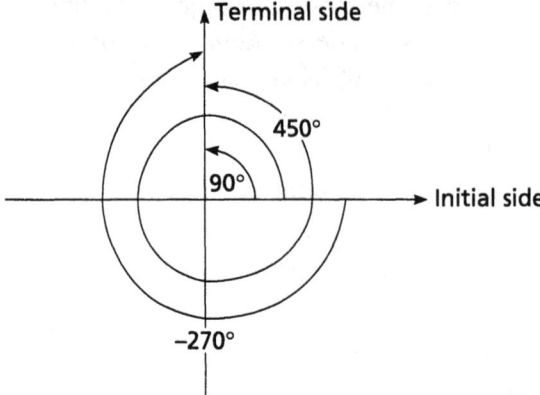

Example 1:

Name and draw four coterminal angles for 60°, two positive and two negative (answers may vary).

Solution:

To find coterminal angles all we have to do is add or subtract 360° or any multiple of 360° to 60°. We'll add 360° and 720° to 60° and subtract 360° and 720° from 60°. Our angles are 60° + 360° = 420°, 60° + 720° = 780°, 60° − 360° = −300°, and 60° − 720° = −660°. The four coterminal angles we used were 420°, 780°, −300°, and −660°; you could have other answers. The angles are shown below.

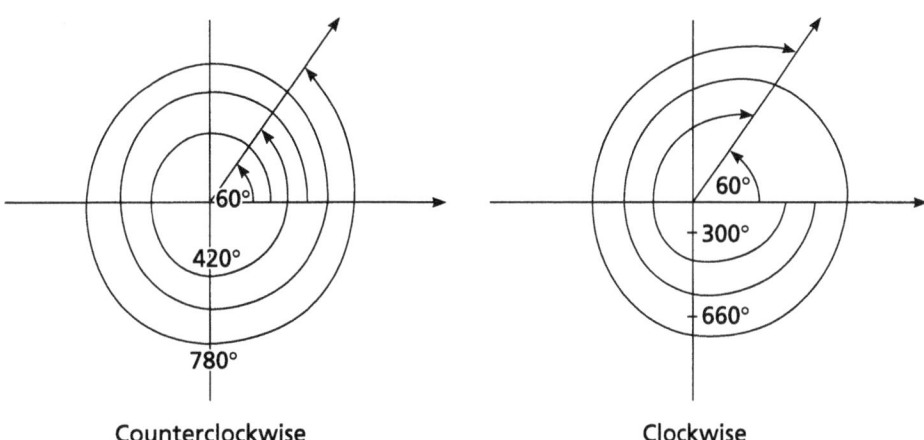

Counterclockwise   Clockwise

Sometimes angles are not represented in degrees, but in a different unit, called *radians*. If a central angle of a circle intercepts an arc equal in length to the radius of the circle, the central angle is defined as one radian.

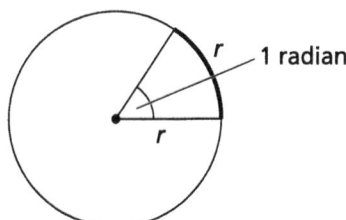

Since the radius can be marked off along the circumference 2π (about 6.283) times, we see that $2\pi = 360°$, or we'll say $\pi = 180°$. If the degree symbol ° isn't written, it's assumed the angle is represented in radians, not degrees. You should always check to see if your calculator is in degree form or radian form. Both degree form and radian form are used, so it's important that we are able to convert from degree form to radian form and from radian form to degree form.

To convert from degree form to radian form, multiply by $\dfrac{\pi}{180}$.

Example 2:

Convert 90° to radian form.

Solution:

$$90° \left(\frac{\pi}{180°} \right) = \frac{\pi}{2}$$

To convert from radian form to degree form, multiply by $\frac{180}{\pi}$.

Example 3:

Convert $\frac{\pi}{2}$ to degrees.

Solution:

$$\frac{\pi}{2} \left(\frac{180}{\pi} \right) = 90°$$

Example 4:

Draw the unit circle from the figure on page 128, but this time label the same angles in radian form. These are standard angles; it's important that you memorize their radian forms.

Solution:

First we have to multiply all the angles by $\frac{\pi}{180}$ to convert to radian form.

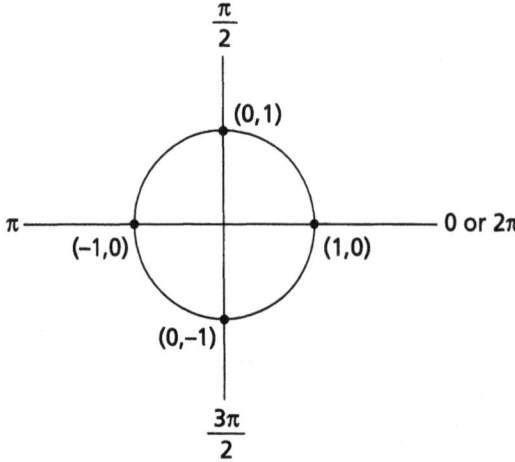

To find a coterminal angle for an angle in radian form, just add or subtract 2π or a multiple of 2π.

Trigonometry

Example 5:
Find a positive and a negative coterminal angle in radian form for $\frac{\pi}{4}$.

Solution:
We'll add and subtract 2π to $\frac{\pi}{4}$.

$$2\pi + \frac{\pi}{4} = \frac{8\pi}{4} + \frac{\pi}{4} = \frac{9\pi}{4} \qquad 2\pi - \frac{\pi}{4} = \frac{8\pi}{4} - \frac{\pi}{4} = \frac{7\pi}{4}$$

SELF-TEST 1:

1. Name two positive coterminal angles for 45°.

2. Name two negative coterminal angles for –120°.

3. Name two negative coterminal angles for $\frac{-4\pi}{5}$.

4. Name two positive coterminal angles for $\frac{\pi}{9}$.

5. Convert the following angles in degrees to radian: 30°, 45°, and 60°, and memorize these forms, since they'll be used a lot.

6. Convert the following angles to degrees: $\frac{5\pi}{4}, \frac{5\pi}{6}, \frac{-5\pi}{2}$, and state the quadrant they're in.

ANSWERS:

1. $45° + 360° = 405°$ $\qquad 45° + 720° = 765°$

2. $-120° - 360° = -480° \qquad -120° - 720° = -840°$

3. $\frac{-4\pi}{5} - 2\pi = \frac{-4\pi}{5} - \frac{10\pi}{5} = \frac{-14\pi}{5} \qquad \frac{-4\pi}{5} - 4\pi = \frac{-4\pi}{5} - \frac{20\pi}{5} = \frac{-24\pi}{5}$

4. $\frac{\pi}{9} + 2\pi = \frac{\pi}{9} + \frac{18\pi}{9} = \frac{19\pi}{9} \qquad \frac{\pi}{9} + 4\pi = \frac{\pi}{9} + \frac{36\pi}{9} = \frac{37\pi}{9}$

5. $30\left(\frac{\pi}{180}\right) = \frac{\pi}{6} \qquad 45\left(\frac{\pi}{180}\right) = \frac{\pi}{4} \qquad 60\left(\frac{\pi}{180}\right) = \frac{\pi}{3}$

6. $\frac{5\pi}{4}\left(\frac{180}{\pi}\right) = 225° \qquad \frac{5\pi}{6}\left(\frac{180}{\pi}\right) = 150° \qquad \frac{-5\pi}{2}\left(\frac{180}{\pi}\right) = -450°$
   III $\qquad\qquad\qquad$ II $\qquad\qquad\qquad$ none

2 Right-Triangle Trigonometry

Our first look at the six trig functions will be through the use of a right triangle. Many applications can be drawn in the form of a right triangle. Right triangles are triangles that have a 90-degree angle. The figure below shows a standard right triangle. Notice that the three sides are labeled according to their relationship to an angle other than the right angle. The side opposite the right angle is always called the hypotenuse. Angles are always labeled by Greek letters or capital letters; sides are always labeled by small letters.

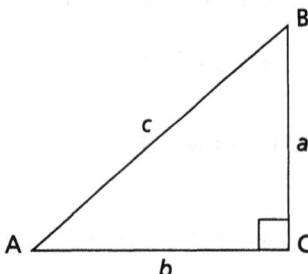

Using the lengths of the sides to form six ratios, we define the six trig functions—sine, cosine, tangent, cosecant, secant, and cotangent—in table 5.1. Their abbreviations are sin, cos, tan, csc, sec, and cot. We assume that θ is an acute angle (0° < θ < 90°) of a right triangle.

Table 5.1 The Six Trigonometric Functions of a Right Triangle

$$\sin \theta = \frac{opp}{hyp} \qquad \csc \theta = \frac{hyp}{opp}$$

$$\cos \theta = \frac{adj}{hyp} \qquad \sec \theta = \frac{hyp}{adj}$$

$$\tan \theta = \frac{\sin}{\cos} = \frac{opp}{adj} \qquad \cot \theta = \frac{\cos}{\sin} = \frac{adj}{opp}$$

Notice that the sin and the csc are reciprocal functions; the same is also true for the cos and the sec, and the tan and the cot.

Example 6:

Find the values of the six trig functions of angles A and B shown here.

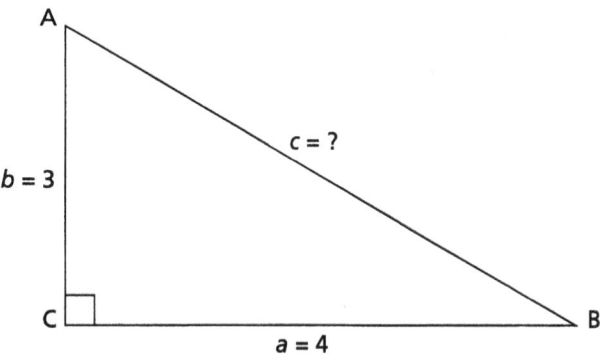

Solution:

Since the measure of side c is not given, we'll start by using the Pythagorean theorem from chapter 1, section 5 to find c.

$a^2 + b^2 = c^2$
$4^2 + 3^2 = c^2$
$16 + 9 = c^2$
$\quad\quad 25 = c^2$
$\quad\quad\; c = 5$

$\sin A = \dfrac{opp}{hyp} = \dfrac{a}{c} = \dfrac{4}{5}$ $\quad \csc A = \dfrac{5}{4}$ $\quad \sin B = \dfrac{b}{c} = \dfrac{3}{5}$ $\quad \csc B = \dfrac{5}{3}$

$\cos A = \dfrac{adj}{hyp} = \dfrac{b}{c} = \dfrac{3}{5}$ $\quad \sec A = \dfrac{5}{3}$ $\quad \cos B = \dfrac{a}{c} = \dfrac{4}{5}$ $\quad \sec B = \dfrac{5}{4}$

$\tan A = \dfrac{opp}{adj} = \dfrac{a}{b} = \dfrac{4}{3}$ $\quad \cot A = \dfrac{3}{4}$ $\quad \tan B = \dfrac{b}{a} = \dfrac{3}{4}$ $\quad \cot B = \dfrac{4}{3}$

SELF-TEST 2: Find the values of the six trig functions of angles A and B for these triangles:

1.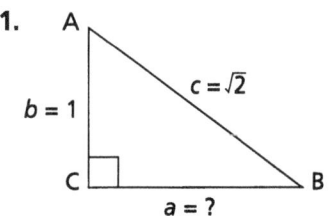

2.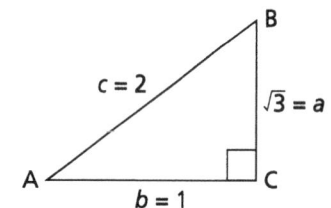

ANSWERS:

$a^2 + b^2 = c^2$

1. $a = \sqrt{c^2 - b^2}$
$a = \sqrt{2 - 1} = 1$

In this case, sides *a* and *b* are equal. When that happens, we can assume that sides AC and AB are equal. That means that angles A and B are equal. As you already know that the sum of the angles of a triangle is 180°, angles A and B would have to be 45° each because angle C is 90°. This type of triangle is called an isosceles triangle. Because angles A and B are both 45°, the trig functions of A and B should be the same.

$\sin A = \dfrac{1}{\sqrt{2}} = \dfrac{\sqrt{2}}{2}$ $\csc A = \dfrac{2}{\sqrt{2}} = \sqrt{2}$ $\sin B = \dfrac{1}{\sqrt{2}} = \dfrac{\sqrt{2}}{2}$ $\csc B = \dfrac{2}{\sqrt{2}} = \sqrt{2}$

$\cos A = \dfrac{1}{\sqrt{2}} = \dfrac{\sqrt{2}}{2}$ $\sec A = \dfrac{2}{\sqrt{2}} = \sqrt{2}$ $\cos B = \dfrac{1}{\sqrt{2}} = \dfrac{\sqrt{2}}{2}$ $\sec B = \dfrac{2}{\sqrt{2}} = \sqrt{2}$

$\tan A = \dfrac{1}{1} = 1$ $\cot A = \dfrac{1}{1} = 1$ $\tan B = \dfrac{1}{1} = 1$ $\cot B = \dfrac{1}{1} = 1$

2. $\sin A = \dfrac{\sqrt{3}}{2}$ $\csc A = \dfrac{2}{\sqrt{3}} = \dfrac{2\sqrt{3}}{3}$ $\sin B = \dfrac{1}{2}$ $\csc B = 2$

$\cos A = \dfrac{1}{2}$ $\sec A = 2$ $\cos B = \dfrac{\sqrt{3}}{2}$ $\sec B = \dfrac{2}{\sqrt{3}} = \dfrac{2\sqrt{3}}{3}$

$\tan A = \sqrt{3}$ $\cot A = \dfrac{1}{\sqrt{3}} = \dfrac{\sqrt{3}}{3}$ $\tan B = \dfrac{1}{\sqrt{3}} = \dfrac{\sqrt{3}}{3}$ $\cot B = \sqrt{3}$

3 Trigonometric Functions of Any Angle

In this section you will learn how to find the values of the six trig functions of any angle. For us to do this, we will have to use reference angles. Reference angles are angles in quadrant I. For us to derive the values of the trig functions of the standard reference angles, you must first understand the relationship between the unit circle and a right triangle. The top figure on page 135 shows the relationship between the sides of a triangle on the unit circle and the six trig functions. It also shows that sin θ is the *y* value of a point on the unit circle, and cos θ is the *x* value of a point on the unit circle. The $\sin \theta = \dfrac{y}{r}$, which yields $y = r \sin \theta$. We know $r = 1$, so we can say $y = \sin \theta$.

We'll say the sin θ is the *y* value. In the same way, $\cos \theta = \dfrac{x}{r}$, which yields $x = \cos \theta$. In both cases the *y* is the distance off the *x*-axis, and the *x* is the distance along the *x*-axis, while *r* is the hypotenuse.

We'll use the bottom figure on page 135 to determine the values of the sin, cos, and tan values of 90°, 180°, and 270°. We'll also label the signs of the trig functions in each quadrant. The figure shows the unit circle with its intercepts labeled. Remember, the *x* values are the values

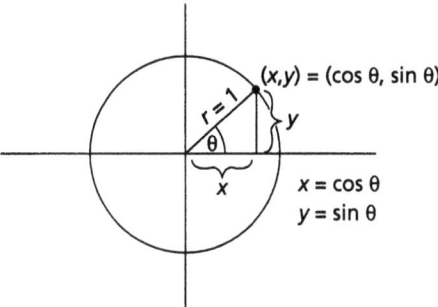

of the cosines of those angles, and the y values are the values of the sines of those angles. And let's not forget that $\frac{y}{x}$ is the value of the tangent of the angles. From this figure we can see that:

sin 0° = 0	cos 0° = 1	tan 0° = 0
sin 90° = 1	cos 90° = 0	tan 90° = undefined
sin 180° = 0	cos 180° = −1	tan 180° = 0
sin 270° = −1	cos 270° = 0	tan 270° = undefined

It also shows us that the sin θ (which is the y value) is positive in quadrants I and II and negative in quadrants III and IV. The cos θ (which is the x value) is positive in quadrants I and IV and negative in quadrants II and III. The tan θ $\left(\text{which is } \frac{y}{x}\right)$ is positive in quadrants I and III and negative in quadrants II and IV.

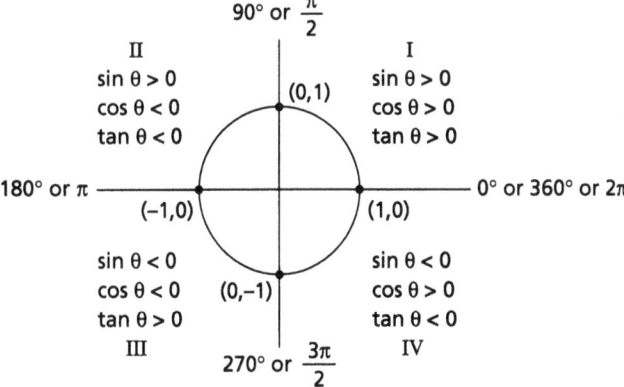

In the previous section, the triangles in self-test 2 were standard triangles, which gave us the standard reference angles in quadrant I. The first one, which is shown below, has two 45° angles.

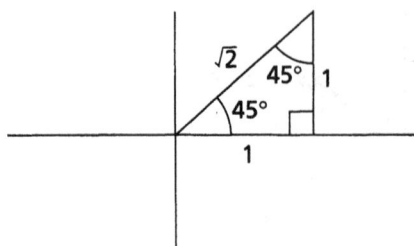

The values of the first three trig functions of a 45° angle were derived from the first problem in self-test 2.

$$\sin 45° = \frac{\sqrt{2}}{2} \qquad \cos 45° = \frac{\sqrt{2}}{2} \qquad \tan 45° = 1$$

The other reference angles we should study are 30° and 60°. Without your knowing it, we've already done this in the second problem in self-test 2. Let's use the triangles below to find the values of the first three trig functions of these angles. Figure a shows a triangle with all sides and angles of equal size. This is called an equilateral triangle. Since we know that the sum of the angles of a triangle is 180° and that all three angles are equal, we know the angles are 60° each. To create a right triangle we'll drop a perpendicular bisector to form a right angle. This is shown in figure b. This cuts one of the 60° angles in half to make two 30° angles. In figure c we'll use one of the right triangles to find the trig functions of 30° and 60°.

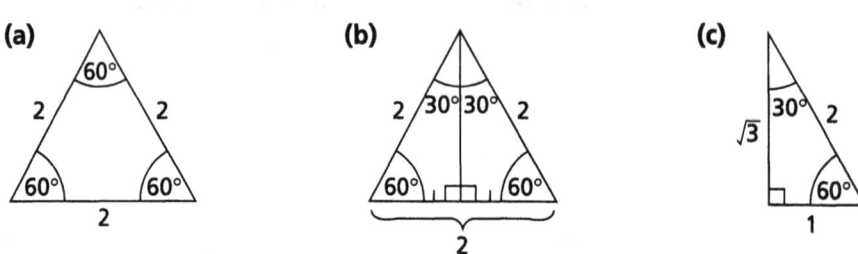

$$\sin 30° = \frac{1}{2}, \text{ which is the same as the } \cos 60°.$$

$$\cos 30° = \frac{\sqrt{3}}{2}, \text{ which is the same as the } \sin 60°.$$

$$\tan 30° = \frac{\sqrt{3}}{3}, \text{ which is } not \text{ the same as the } \tan 60° = \sqrt{3}.$$

Table 5.2 summarizes the trig functions of the basic angles we will use most often. You should memorize these in degree and radian form.

Table 5.2 Trigonometric Functions of Basic Angles

Degrees	0° or 360°	30°	45°	60°	90°	180°	270°
Radians	0 or 2π	$\dfrac{\pi}{6}$	$\dfrac{\pi}{4}$	$\dfrac{\pi}{3}$	$\dfrac{\pi}{2}$	π	$\dfrac{3\pi}{2}$
sin	0	$\dfrac{1}{2}$	$\dfrac{\sqrt{2}}{2}$	$\dfrac{\sqrt{3}}{2}$	1	0	−1
cos	1	$\dfrac{\sqrt{3}}{2}$	$\dfrac{\sqrt{2}}{2}$	$\dfrac{1}{2}$	0	−1	0
tan	0	$\dfrac{\sqrt{3}}{3}$	1	$\sqrt{3}$	undefined	0	undefined

Let's try a few problems using reference angles to find the exact value of angles that are not in quadrant I. The reference angles we'll use are 30° or $\dfrac{\pi}{6}$, 45° or $\dfrac{\pi}{4}$, and 60° or $\dfrac{\pi}{3}$.

Example 7:

Find the exact value of cos 120°.

Solution:

The figure below shows 120°, which is in quadrant II. To find its corresponding reference angle, we measure how far off the x-axis 120° is in quadrant II. To find how far off the x-axis it is, we subtracted 180° − 120° = 60°, so 60° is its reference angle. The cos of 60° is $\dfrac{1}{2}$, but the cos 120° is $-\dfrac{1}{2}$ because the x values in quadrant II are negative.

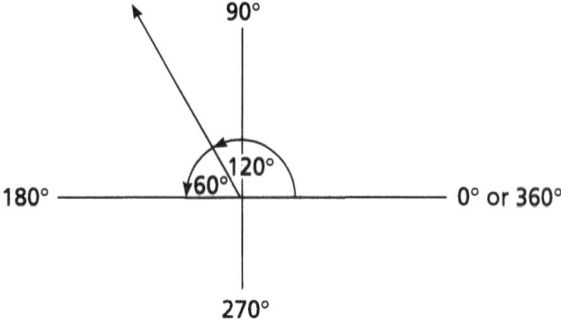

Example 8:
Find the exact value of tan 210°.

Solution:
The figure below shows 210°, which is in quadrant III. To find its corresponding reference angle, we measure how far off the *x*-axis 210° is. To find how far off the *x*-axis it is, we subtract 210° − 180° = 30°, so 30° is its reference angle. The tan of 30° is $\frac{\sqrt{3}}{3}$. The tangent in quadrant III is positive because $\frac{y}{x}$ in quadrant III would be a negative divided by a negative, which is a positive, so tan 210° is $\frac{\sqrt{3}}{3}$.

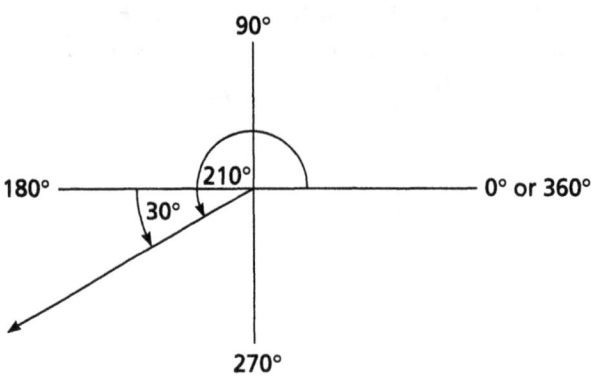

Example 9:
Find the exact value of sin 315°.

Solution:
The figure below shows 315°, which is in quadrant IV. To find its corresponding reference angle, we measure how far off the *x*-axis 315° is. To find how far off the *x*-axis it is, we subtract 360° − 315° = 45°, making 45° its reference angle. The sin of 45° is $\frac{\sqrt{2}}{2}$. Because *y* values are negative in quadrant IV, sin 315° is $-\frac{\sqrt{2}}{2}$.

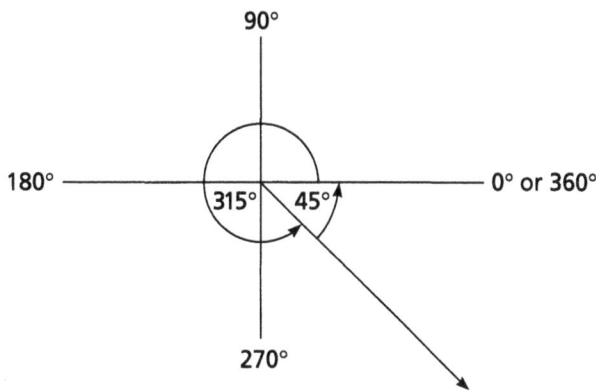

Let's summarize how to find reference angles in quadrants II, III, and IV.

	In degrees	In radians
Quadrant II	$180° - \theta$	$\pi - \theta$
Quadrant III	$\theta - 180°$	$\theta - \pi$
Quadrant IV	$360° - \theta$	$2\pi - \theta$

Example 10:
Find the exact value of $\sin \dfrac{5\pi}{6}$.

Solution:
$\dfrac{5\pi}{6}$ is in quadrant II, where the sin is positive. Its reference angle is $\pi - \dfrac{5\pi}{6} = \dfrac{6\pi}{6} - \dfrac{5\pi}{6} = \dfrac{1\pi}{6}$. The $\sin \dfrac{\pi}{6} = \dfrac{1}{2}$, so $\sin \dfrac{5\pi}{6} = \dfrac{1}{2}$.

SELF-TEST 3:

Find the exact values of the trig functions of the following angles:

1. $\tan 225°$
2. $\cos 150°$
3. $\tan 300°$
4. $\sin 135°$
5. $\sin \dfrac{5\pi}{4}$
6. $\cos \dfrac{2\pi}{3}$
7. $\tan \dfrac{7\pi}{4}$
8. $\cos \dfrac{11\pi}{6}$

ANSWERS:

1. 225° is in quadrant III, where the tan is positive. Its reference angle is found by subtracting $225° - 180° = 45°$. Tan 45° = 1, so tan 225° = 1.

2. 150° is in quadrant II, where the cos is negative. Its reference angle is $180° - 150° = 30°$. The $\cos 30° = \dfrac{\sqrt{3}}{2}$, so $\cos 150° = -\dfrac{\sqrt{3}}{2}$.

3. 300° is in quadrant IV, where the tan is negative. Its reference angle is 360° − 300° = 60°. The tan 60° = $\sqrt{3}$, so tan 300° = −$\sqrt{3}$.

4. 135° is in quadrant II, where the sin is positive. Its reference angle is 180° − 135° = 45°. The sin of 45° = $\frac{\sqrt{2}}{2}$, so sin 135° = $\frac{\sqrt{2}}{2}$.

5. $\frac{5\pi}{4}$ is in quadrant III, where the sin is negative. Its reference angle is $\frac{5\pi}{4} - \frac{4\pi}{4} = \frac{\pi}{4}$. The sin $\frac{\pi}{4} = \frac{\sqrt{2}}{2}$, so sin $\frac{5\pi}{4} = -\frac{\sqrt{2}}{2}$.

6. $\frac{2\pi}{3}$ is in quadrant II, where the cos is negative. Its reference angle is $\pi - \frac{2\pi}{3} = \frac{\pi}{3}$. The cos $\frac{\pi}{3} = \frac{\sqrt{1}}{2}$, so cos $\frac{2\pi}{3} = -\frac{\sqrt{1}}{2}$.

7. $\frac{7\pi}{4}$ is in quadrant IV, where the tan is negative. Its reference angle is $2\pi - \frac{7\pi}{4} = \frac{8\pi}{4} - \frac{7\pi}{4} = \frac{\pi}{4}$. The tan $\frac{\pi}{4} = 1$, so tan $\frac{7\pi}{4} = -1$.

8. $\frac{11\pi}{6}$ is in quadrant IV, where the cos is positive. Its reference angle is $2\pi - \frac{11\pi}{6} = \frac{12\pi}{6} - \frac{11\pi}{6} = \frac{\pi}{6}$. The cos $\frac{\pi}{6} = \frac{\sqrt{3}}{2}$, so cos $\frac{11\pi}{6} = \frac{\sqrt{3}}{2}$.

4 Graphs of the Basic Trigonometric Functions

In this section we will show you the graphs of the basic trig functions. We will not go into how to draw these graphs by hand. The use of graphics calculators can save us from hand-drawing every graph. To use graphics calculators you must be able to set the domain and range of the calculators. As we show you the following graphs, we want you to pay close attention to the domain and range of the functions, their minimum and maximum values, and their intercepts. Let's start by using the values in table 5.3 to sketch the graphs of $y = \sin x$ and $y = \cos x$ on the Cartesian coordinate system.

Table 5.3

x	sin x	cos x
0	0	1
$\frac{\pi}{2}$	1	0
π	0	−1
$\frac{3\pi}{2}$	−1	0
2π	0	1

Below are shown the graphs of $y = \sin x$ and $y = \cos x$. The domain is all real numbers, and the range is from −1 to 1. Notice that they both have the same form except one is shifted a little to the right of the other. They both repeat themselves every 2π units. The distance along the x-axis it takes for the curve to repeat itself is called a period. Both the sin x and cos x functions have a period of 2π units. The maximum height of the graph is called the amplitude. The amplitude of $y = \sin x$ and $y = \cos x$ is 1.

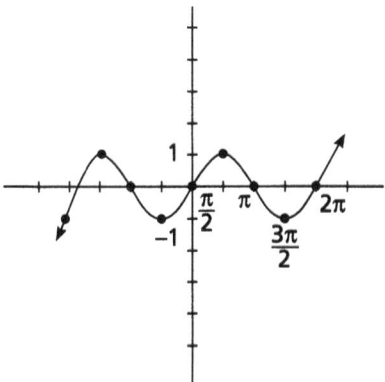

$y = \sin x$
Domain: all real numbers
Range: [−1,1]

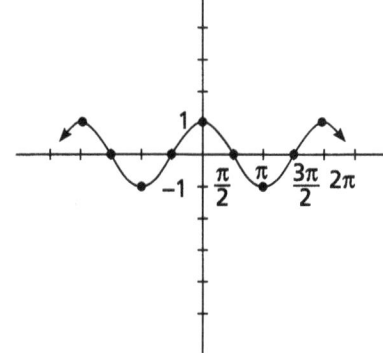

$y = \cos x$
Domain: all real numbers
Range: [−1,1]

Now that you know the form of the graphs of $y = \sin x$ and $y = \cos x$, all you need to learn is the graphs of the remaining trig functions. Table 5.4 shows the values for $y = \csc x$, $y = \sec x$, $y = \tan x$, and $y = \cot x$.

Table 5.4

x	csc x	sec x	tan x	cot x
0 or 2π	undef.	1	0	undef.
$\dfrac{\pi}{6}$	2	$\dfrac{2\sqrt{3}}{3}$	$\dfrac{\sqrt{3}}{3}$	$\sqrt{3}$
$\dfrac{\pi}{4}$	$\sqrt{2}$	$\sqrt{2}$	1	1
$\dfrac{\pi}{3}$	$\dfrac{2\sqrt{3}}{3}$	2	$\sqrt{3}$	$\dfrac{\sqrt{3}}{3}$
$\dfrac{\pi}{2}$	1	undef.	undef.	0
π	undef.	−1	0	undef.
$\dfrac{3\pi}{2}$	−1	undef.	undef.	0

The figures below show the graphs of the remaining trig functions with their domains and ranges.

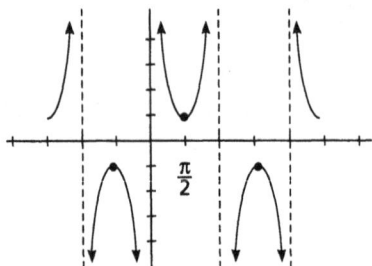

$y = \csc x$

Domain: all $x \neq n\pi$

Range: $(-\infty, -1]$ and $[1, \infty)$

Vertical asymptotes: $x = n\pi$

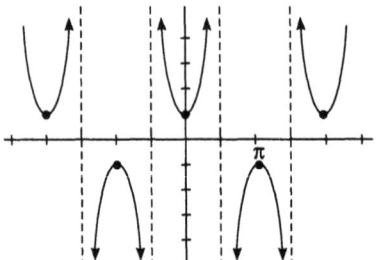

$y = \sec x$

Domain: all $x \neq \frac{\pi}{2} + n\pi$

Range: $(-\infty, -1]$ and $[1, \infty)$

Vertical asymptotes: $x = \frac{\pi}{2} + n\pi$

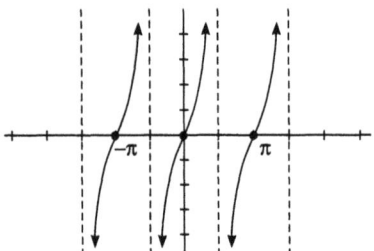

$y = \tan x$

Domain: all $x \neq \frac{\pi}{2} + n\pi$

Range: $(-\infty, \infty)$

Vertical asymptotes: $x = \frac{\pi}{2} + n\pi$

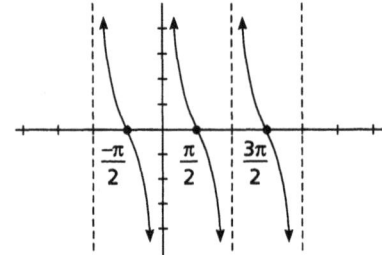

$y = \cot x$

Domain: all $x \neq n\pi$

Range: $(-\infty, \infty)$

Vertical asymptotes: $x = n\pi$

5 Inverse Trigonometric Functions

In chapter 2, section 4 we discussed inverse functions. The inverse of addition is subtraction, the inverse of multiplication is division, the inverse of an exponential function is its corresponding logarithmic function. Trig functions also have inverse functions. Earlier in this chapter, when given an angle we were asked to find the value of the trig functions of that angle. In this chapter we'll invert the process: We'll give you the value of a trig function and ask you to tell us what the angle is.

For example, earlier we asked you to find the value of the sin of 30°, which, as you no doubt know, is $\frac{1}{2}$. In this section we'll ask you to find the angle or angles whose sin is $\frac{1}{2}$. Instead of saying, "What angle or angles have a sin of $\frac{1}{2}$?," we'll write this question as $\sin^{-1} \frac{1}{2}$. This is read, "the sine inverse of $\frac{1}{2}$." The −1 over the sin does *not* mean that the sin is raised to the negative first power; it means sin inverse. (Another way of stating the same thing is $\arcsin \frac{1}{2}$. This is read, "the arcsin of $\frac{1}{2}$." We will be using $\sin^{-1}$ notation here.) When we're trying to find an angle, we use inverse functions. Here are the graphs of the three basic trig inverse functions. We would like you to memorize the domains and ranges of the inverse functions shown below. If a value is not in the domain of the function, its inverse does not exist.

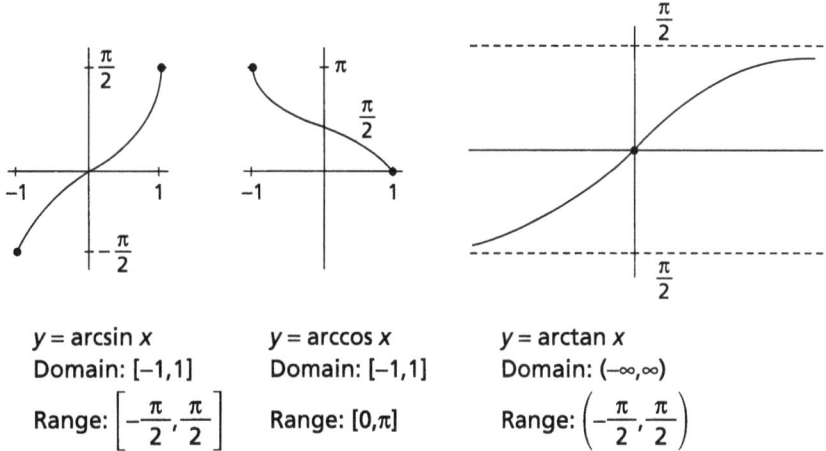

$y = \arcsin x$
Domain: $[-1, 1]$
Range: $\left[-\frac{\pi}{2}, \frac{\pi}{2}\right]$

$y = \arccos x$
Domain: $[-1, 1]$
Range: $[0, \pi]$

$y = \arctan x$
Domain: $(-\infty, \infty)$
Range: $\left(-\frac{\pi}{2}, \frac{\pi}{2}\right)$

For all the problems in this section we'll state the angles in degree form and in radian form. It's important to get used to using both forms.

Example 11:
Find the $\sin^{-1} 1$.

Solution:
This problem asks us to find the angle whose sin is 1. The answer is 90° or $\frac{\pi}{2}$.

Example 12:
Find the $\cos^{-1} \frac{\sqrt{2}}{2}$.

Solution:
This problem asks us to find the angle(s) whose cos is $\frac{\sqrt{2}}{2}$. The cos is positive in quadrants I and IV, so we're looking for two angles. The angle in quadrant I is $\frac{\pi}{4}$ or 45° (see table 5.2). To find the angle in quadrant IV we also have to move 45° off the x-axis. We'll do this by subtracting: 360° − 45° = 315°, or $\frac{7\pi}{4}$.

Example 13:
Find the $\tan^{-1} -\frac{\sqrt{3}}{3}$.

Solution:
This problem asks us to find the angle(s) whose tan is $-\frac{\sqrt{3}}{3}$. The tangent is negative in quadrants II and IV. The reference angle in quadrant I whose tangent is $\frac{\sqrt{3}}{3}$ is 30°. To find the angle in quadrant II we move 30° off the x-axis by subtracting: 180° − 30° = 150°, or $\frac{5\pi}{6}$. To find the angle in quadrant IV we move 30° off the x-axis by subtracting: 360° − 30° = 330°, or $\frac{11\pi}{6}$.

Example 14:
Find the $\sin^{-1} -\frac{\sqrt{2}}{2}$.

Solution:
The sin is negative in quadrants III and IV. The reference angle in quadrant I whose sin is $\frac{\sqrt{2}}{2}$ is 45°. To find the angle in quadrant III we move 45° off the x-axis by adding: 180° + 45° = 225°, or $\frac{5\pi}{4}$. To find the angle in quadrant IV we move 45° off the x-axis by subtracting: 360° − 45° = 315°, or $\frac{7\pi}{4}$.

SELF-TEST 4: Find the angle in degree form and in radian form for the following problems:

1. $\sin^{-1} \dfrac{1}{2}$
2. $\cos^{-1} -\dfrac{\sqrt{3}}{2}$
3. $\tan -\sqrt{3}$
4. $\tan^{-1} 0$
5. $\sin^{-1} \dfrac{\sqrt{2}}{2}$
6. $\cos^{-1} -\dfrac{1}{2}$

ANSWERS:

1. Sin is positive in quadrants I and II. The reference angle whose sin is $\dfrac{1}{2}$ is 30°, or $\dfrac{\pi}{6}$. To find the angle in quadrant II we move 30° off the x-axis by subtracting: 180° − 30° = 150°, or $\dfrac{5\pi}{6}$.

2. Cos is negative in quadrants II and III. The reference angle whose cos is $\dfrac{\sqrt{3}}{2}$ is 30°, or $\dfrac{\pi}{6}$. To find the angle in quadrant II we move 30° off the x-axis by subtracting: 180° − 30° = 150°, or $\dfrac{5\pi}{6}$. To find the angle in quadrant III we move 30° off the x-axis by adding: 180° + 30° = 210°, or $\dfrac{7\pi}{6}$.

3. The tan is negative in quadrants II and IV. The reference angle whose tan is $\sqrt{3}$ is 60°, or $\dfrac{\pi}{3}$. To find the angle in quadrant II we move 60° off the x-axis by subtracting: 180° − 60° = 120°, or $\dfrac{2\pi}{3}$. To find the angle in quadrant IV we move 60° off the x-axis by subtracting: 360° − 60° = 300°, or $\dfrac{5\pi}{3}$.

4. $\tan^{-1} = 0$. Remember, at $\tan = \dfrac{y}{x}$ the only time it can be 0 is when y is 0. The angles are 180°, or π and 0°.

5. $\sin^{-1} \dfrac{\sqrt{2}}{2}$. The sin is positive in quadrants I and II. The reference angle whose sin is $\dfrac{\sqrt{2}}{2}$ is 45°, or $\dfrac{\pi}{4}$. To find the angle in quadrant II we move 45° off the x-axis by subtracting: 180° − 45° = 135°, or $\dfrac{3\pi}{4}$.

6. $\cos^{-1} -\dfrac{1}{2}$. The cos is negative in quadrants II and III. The reference angle whose cos is $\dfrac{1}{2}$ is 60°, or $\dfrac{\pi}{3}$. To find the angle in quadrant II we move 60° off the x-axis by subtracting: 180° − 60° = 120°, or $\dfrac{2\pi}{3}$. To find the angle in quadrant III we move 60° off the x-axis by adding: 180° + 60° = 240°, or $\dfrac{4\pi}{3}$.

6 Applications

In this section we'll apply everything we've learned so far in this chapter to real-life word problems. Before we can do that, we should look at a few examples on how to solve right triangles. "Solving a triangle" means finding the measure of all its sides and angles.

Example 15:

Solve the following right triangle:

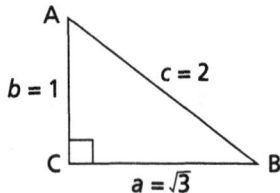

Solution:

We already know the measure of all the sides: $a = \sqrt{3}$, $b = 1$, and $c = 2$. We know that $C = 90$ and that the sum of A and B must be 90. It doesn't matter whether we find A first or B first. Let's find B. To find the measure of an angle we have to use inverse functions, and it doesn't matter which inverse function. Let's use $\cos^{-1}$. Angle $B = \cos^{-1} \frac{\sqrt{3}}{2}$. You may recall that the $\cos 30°$ is $\frac{\sqrt{3}}{2}$, so $B = 30°$. Angle A is found by subtracting $90° - 30° = 60°$. Now the triangle is solved.

A = 60°   B = 30°   C = 90° $a = \sqrt{3}$ $b = 1$ $c = 2$

Example 16:

Solve the following right triangle:

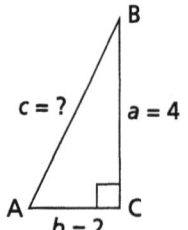

Solution:

To find the measure of side *c* we'll use the Pythagorean theorem:

$c^2 = a^2 + b^2$
$c^2 = 4^2 + 2^2$
$c^2 = 20$
$c = \sqrt{20}$
$c = 2\sqrt{5}$

Now we have to find the measure of the angles of the given triangle. We already know that angle C is 90°. The sum of angles A and B must add up to 90°. To find the measure of one of these angles we must use inverse functions. We can use A to find B or B to find A; it doesn't matter which one we start with. Let's use A. We have to chose an inverse trig function to use; it doesn't matter which one. Let's use $\sin^{-1}$. Using the information given in the triangle above, the $\sin^{-1} A = \dfrac{4}{2\sqrt{5}}$. Based on our work in the previous pages, we know that $\dfrac{4}{2\sqrt{5}}$ will not give us a standard angle, so we have to use our calculator to find the $\sin^{-1} \dfrac{4}{2\sqrt{5}}$. This gives us an angle of approximately 63°. Once we know the value of A, finding B is simply a matter of subtracting: 90° − 63° = 27°. Now the triangle is solved.

A = 63°   B = 27°   C = 90°   a = 3   b = 4   c = 2√5

Example 17:
Solve the following triangle:

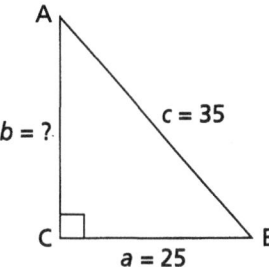

Solution:

We know the measures of sides *a* and *c*, so our first step will be to use the Pythagorean theorem to find the measure of side *b*.

$$c^2 = a^2 + b^2$$
$$35^2 = 25^2 + b^2$$
$$b = \sqrt{1225 - 625}$$
$$b = 10\sqrt{6}$$

Let's start by finding the measure of angle A. $A = \sin^{-1}\dfrac{25}{35} \approx 46°$. $B = 90° - 46° = 44°$. The triangle is solved.

A = 46°   B = 44°   C = 90°   a = 25   b = 10√6   c = 35

Now it's your turn to solve a few triangles.

SELF-TEST 5: Use the triangle below to solve the triangles with the following measures. Round your answers to two decimal places.

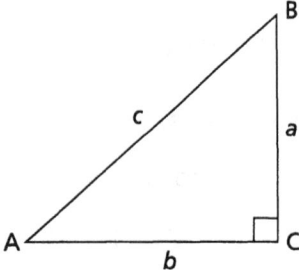

1. $A = 20°, b = 10$ **2.** $a = 6, b = 10$ **3.** $B = 74°, a = 46$

ANSWERS:

1. A = 20°, B = 70° C = 90°
$$\tan A = \frac{a}{10}$$
$a = 10 \tan 20° \approx 3.64$
$c^2 = 3.64^2 + 10^2$
$c \approx 10.64$

2. $a = 6, b = 10$
$$B = \tan^{-1}\frac{10}{6}$$
$B \approx 59.04°$
$A = 90° - 59.04° = 30.96°$
$c^2 = 36 + 100$
$c \approx 11.66$

3. B = 74°, a = 46, A = 16°, C = 90°
$$\tan 74° = \frac{b}{46}$$
$b = 46 \tan 74° \approx 160.42$
$c^2 = 46^2 + 160.42^2$
$c \approx 166.89$

The following examples are word problems involving right triangles. You can solve the triangles the same way you just solved the previous self-test problems, but we recommend you try to draw each triangle first. It's much easier to solve a word problem if you can visualize the shapes involved.

Example 18:

A surveyor wants to measure how wide a gorge is from point C to point A (see the figure below). She sets up a transit (a surveying instrument used to measure angles) at point C and takes a sighting of point A on the other side of the gorge. After turning through an angle of 90° at C, she paces off a distance of 100 yards to point B. Then, using the transit at B, she finds angle B to be 30°. What is the width of the gorge?

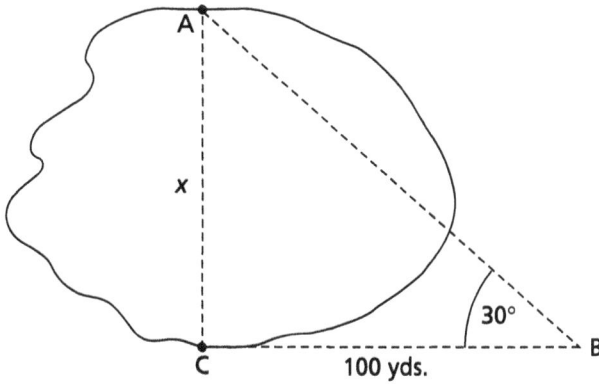

Solution:

In this problem we're looking for the measure of the side opposite 30 and we know the measure of the adjacent side. Since opposite and adjacent are involved, we'll use the tangent function to solve for the unknown side, x.

$$\tan 30° = \frac{x}{100}$$
$$x = 100 \tan 30°$$
$$x = \frac{100\sqrt{3}}{3} \approx 58$$

Example 19:

Heightstown has an ordinance that limits structures to a maximum of 125 feet. When a new building is completed, a surveyor is sent to determine if it is no more than 125 feet tall. Setting up her transit 125 feet away from the building, the surveyor takes an observation on the top of the building and finds the angle of elevation to be 44°. How tall is the building? The angle of elevation is the angle formed with the horizon when looking up.

Solution:

We know the side adjacent to the angle and we want to find the side opposite the angle, so we'll use the tan again.

$$\tan 44° = \frac{x}{125}$$

$x = 125 \tan 44° \approx 121$ ft

This is just under the limit.

SELF-TEST 6:

1. Find the distance across the pond shown below.

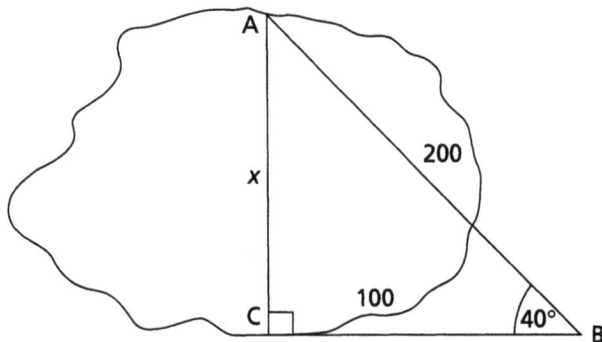

2. A ship just off the coast of Cape Cod has a sighting of a lighthouse that is 182 feet above sea level. If the angle of elevation (looking up) to the top of the lighthouse is 17°, how far is the ship from the base of the lighthouse?

3. A 27-foot ladder leaning against a building makes a 67° angle with the ground. How far up the building does the ladder reach?

4. A straight road with a uniform inclination of 9° leads from a house with an elevation of 5,400 feet to a barn with an elevation of 6,100 feet. What is the length of the road?

5. A guy wire 176 feet long is attached to the top of a radio transmission tower 129 feet high. What is the wire's angle of elevation?

6. What is the angle of depression (the angle formed with the horizon when looking down) formed by the top of a 175-foot tower and a 268-foot supporting guy wire?

7. A straight road 300 feet long leads down from a house on a hill to a garage at the foot of the hill. If the elevation of the house is 2,900 feet and the elevation of the garage is 2,800 feet, find the angle of depression.

Trigonometry   151

ANSWERS:

1. $\sin 40° = \dfrac{x}{200}$

$x = 200 \sin 40° \approx 129$ ft

2. $\tan 17° = \dfrac{182}{x}$

$x \tan 17° = 182$

$x = \dfrac{182}{\tan 17°} \approx 595.3$ ft

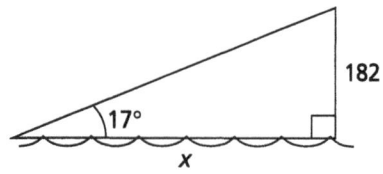

3. $\sin 67° = \dfrac{x}{27}$

$x = 27 \sin 67° \approx 24.85$

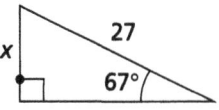

4. $\sin 9° = \dfrac{700'}{x}$

$x = \dfrac{700'}{\sin 9°}$

$x = 4474.72$ ft

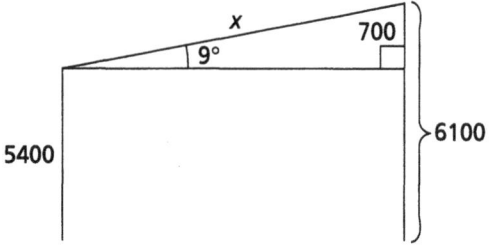

5. $\sin^{-1} \dfrac{129}{176} \approx 47.13$

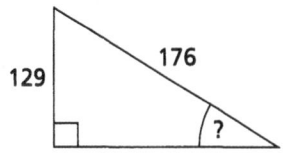

6. $\cos^{-1} \dfrac{175}{268} \approx 49.23$

$x = 90 - 49.23 = 40.77$

7. $\cos^{-1} \dfrac{100}{300} \approx 71$

$90 - 71 = 19$

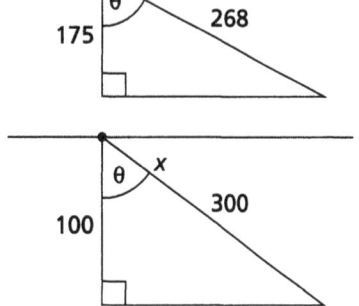

6 Analytic Trigonometry

Sometimes when working on trigonometric problems it's possible to simplify a problem using trig identities. Identities are basic relationships that pop up repeatedly in trigonometric problems. Once a problem has been written in a simpler form it's easier to find the solution to the problem. In this chapter you'll learn:

- the basic trig identities
- how to prove trig identities
- how to solve trig equations
- some basic trig formulas

1 Using Fundamental Identities

Table 6.1 lists some of the basic trig identities. In the following examples we'll use these basic identities to simplify trig expressions. After we've done a few, you'll find they're like solving puzzles: you just have to fit the pieces together.

Analytic Trigonometry

Table 6.1 Fundamental Trigonometric Identities

$\sin u = \dfrac{1}{\csc u}$	$\cos u = \dfrac{1}{\sec u}$	$\tan u = \dfrac{1}{\cot u}$
$\csc u = \dfrac{1}{\sin u}$	$\sec u = \dfrac{1}{\cos u}$	$\cot u = \dfrac{1}{\tan u}$
$\tan u = \dfrac{\sin u}{\cos u}$	$\cot u = \dfrac{\cos u}{\sin u}$	
$\sin^2 u + \cos^2 u = 1$	$1 + \tan^2 u = \sec^2 u$	$1 + \cot^2 u = \csc^2 u$
$1 - \sin^2 u = \cos^2 u$	$1 = \sec^2 u - \tan^2 u$	$1 = \csc^2 u - \cot^2 u$
$1 - \cos^2 u = \sin^2 u$	$\tan^2 u = \sec^2 u - 1$	$\cot^2 u = \csc^2 u - 1$

Example 1:
Simplify $\sin x - \sin x \cos^2 x$.

Solution:
Both terms have a common factor of $\sin x$, so we'll start by factoring out the $\sin x$.

$\sin x - \sin x \cos^2 x$ — Factor out the $\sin x$.
$\sin x(1 - \cos^2 x)$ — Replace $1 - \cos^2 x$ by $\sin^2 x$ (see table 6.1).
$\sin x(\sin^2 x)$
$\sin^3 x$

Example 2:
Simplify $\sin x \sec x$.

Solution:

$\sin x \sec x$ — Replace $\sec x$ by $\dfrac{1}{\cos x}$.

$\sin x\left(\dfrac{1}{\cos x}\right)$

$\dfrac{\sin x}{\cos x} = \tan x$

Example 3:
Simplify $\tan x \csc x$.

Solution:

$\tan x \csc x$ — Replace $\tan x$ by $\dfrac{\sin x}{\cos x}$, and $\csc x$ by $\dfrac{1}{\sin x}$.

$\dfrac{\sin x}{\cos x} \cdot \dfrac{1}{\sin x}$ — Eliminate the common factor of $\sin x$.

$\dfrac{1}{\cos x} = \sec x$ — Replace $\dfrac{1}{\cos x}$ by $\sec x$.

Example 4:
Simplify $\tan^2 x \cos^2 x + \cot^2 x \sin^2 x$.

Solution:

There is no common factor.

$\tan^2 x \cos^2 x + \cot^2 x \sin^2 x$ — We'll begin by substituting some basic identities.

$\dfrac{\sin^2 x}{\cos^2 x} \cdot \cos^2 x + \dfrac{\cos^2 x}{\sin^2 x} \cdot \sin^2 x$ — Reduce the common factors of $\sin^2 x$ and $\cos^2 x$.

$\sin^2 x + \cos^2 x$
$\quad 1$ — Replace $\sin^2 x + \cos^2 x$ by 1.

Example 5:
Rewrite $\dfrac{1}{1 + \sin x}$ so that it is not in fractional form.

Solution:

Nothing factors; there is no basic trig function we can substitute to simplify this expression as it is. The denominator of the fraction is a binomial; because there aren't any other options available to us, we'll begin by multiplying the numerator and the denominator by the conjugate of the denominator. This is called rationalizing the denominator. The conjugate of the denominator is the same binomial, only the sign in the middle is reversed. If we multiply the numerator and the denominator by the same value, it's like multiplying by 1, so multiplying by the conjugate doesn't change the value of the fraction.

$\left(\dfrac{1}{1 + \sin x}\right)\left(\dfrac{1 - \sin x}{1 - \sin x}\right)$ — Next, we'll multiply.

$$\frac{1-\sin x}{1-\sin^2 x}$$ Substitute $\cos^2 x$ for $1 - \sin^2 x$.

$$\frac{1-\sin x}{\cos^2 x}$$ Write this as separate fractions.

$$\frac{1}{\cos^2 x} - \frac{\sin x}{\cos^2 x}$$ Replace $\frac{1}{\cos^2 x}$ by $\sec^2 x$ and separate $\frac{\sin x}{\cos^2 x}$.

$$\sec^2 x - \frac{\sin x}{\cos x} \cdot \frac{1}{\cos x}$$ Replace $\frac{\sin x}{\cos x}$ by $\tan x$ and $\frac{1}{\cos x}$ by $\sec x$.

$$\sec^2 x - \tan x \sec x$$

SELF-TEST 1: Simplify the following trig expressions as much as possible:

1. $\cos^2 x(1 + \tan^2 x)$
2. $\dfrac{\cot x}{\csc x}$
3. $(1 - \cos^2 x)(\csc x)$

4. $\sec^4 x - \tan^4 x$
5. $\dfrac{1}{\tan^2 x + 1}$
6. $\dfrac{1}{1+\cos x} + \dfrac{1}{1 - \cos x}$

7. $\tan x - \dfrac{\sec^2 x}{\tan x}$

ANSWERS:

1. $\cos^2(1 + \tan^2 x)$
   $\cos^2(\sec^2 x)$
   $\cos^2 x\left(\dfrac{1}{\cos^2 x}\right)$
   1

2. $\dfrac{\cot x}{\csc x}$
   $\dfrac{\cos x}{\sin x}$
   $\dfrac{1}{\sin x}$
   $\dfrac{\cos x}{\sin x} \cdot \dfrac{\sin x}{1}$
   $\cos x$

3. $(1 - \cos^2 x)(\csc x)$
   $\sin^2 x\left(\dfrac{1}{\sin x}\right)$
   $\sin x$

4. $\sec^4 x - \tan^4 x$
   $(\sec^2 x - \tan^2 x)(\sec^2 x + \tan^2 x)$
   $1(\sec^2 x + \tan^2 x)$
   $\sec^2 x + \tan^2 x$

5. $\dfrac{1}{\tan^2 x + 1}$
   $\dfrac{1}{\sec^2 x}$
   $\cos^2 x$

6. $\dfrac{1}{1 + \cos x} + \dfrac{1}{1 - \cos x}$
   $\dfrac{1 - \cos x + 1 + \cos x}{(1 + \cos x)(1 - \cos x)}$
   $\dfrac{2}{1 - \cos^2 x}$
   $\dfrac{2}{\sin^2 x}$
   $2\csc^2 x$

7. $\tan x - \dfrac{\sec^2 x}{\tan x}$
   $\dfrac{\tan^2 x - \sec^2 x}{\tan x}$
   $\dfrac{-(\sec^2 x - \tan^2 x)}{\tan x}$
   $-\dfrac{1}{\tan x}$
   $-\cot x$

2 Verifying Trigonometric Identities

In this section we'll use the trig identities from the previous section to prove trigonometric statements. When we're asked to prove trig statements we have to start with one side of the statement and simplify it until it's the same as the other side of the statement.

Example 6:

Prove $\dfrac{\cot x}{\cos x} = \csc x$.

Solution:

We'll start with the left side of the equation and simplify it until it matches the right side.

$\dfrac{\cot x}{\cos x} = \csc x$ Replace $\cot x$ by $\dfrac{\cos x}{\sin x}$.

$\dfrac{\frac{\cos x}{\sin x}}{\frac{\cos x}{1}}$

$\dfrac{\cos x}{\sin x} \cdot \dfrac{1}{\cos x}$ Reduce out the common factor of $\cos x$.

$\dfrac{1}{\sin x}$

$\csc x = \csc x$

Example 7:

Prove $\tan x(\cot x + \tan x) = \sec^2 x$.

Solution:

We'll start by distributing the $\tan x$.

$\tan x(\cot x + \tan x) = \sec^2 x$

$\tan x \cot x + \tan^2 x = \sec^2 x$ Replace $\cot x$ by $\dfrac{1}{\tan x}$.

$\tan x \left(\dfrac{1}{\tan x} \right) + \tan^2 x = \sec^2 x$ Reduce out the common factor of $\tan x$.

$1 + \tan^2 x = \sec^2 x$

$\sec^2 x = \sec^2 x$

Analytic Trigonometry

Example 8:

Prove $\dfrac{\sin x}{1 - \cos x} = \csc x + \cot x$.

Solution:

We're going to show you two methods to prove this statement. The first is to start with the left side of the equation; the second is to start with the right side of the equation.

First method:

$\dfrac{\sin x}{1 - \cos x} = \csc x + \cot x$ We'll start by rationalizing the denominator.

$\left(\dfrac{\sin x}{1 - \cos x}\right)\left(\dfrac{1 + \cos x}{1 + \cos x}\right) = \csc x + \cot x$ Now multiply the denominators.

$\dfrac{\sin x(1 + \cos x)}{1 - \cos^2 x}$ Replace $1 - \cos^2 x$ with $\sin^2 x$.

$\dfrac{\sin x(1 + \cos x)}{\sin^2 x} = \csc x + \cot x$ Reduce the common factor of $\sin x$.

$\dfrac{1 + \cos x}{\sin x} = \csc x + \cot x$ Write the expression as separate fractions.

$\dfrac{1}{\sin x} + \dfrac{\cos x}{\sin x} = \csc x + \cot x$ Use identities.

$\csc x + \cot x = \csc x + \cot x$

Second method:

$\dfrac{\sin x}{1 - \cos x} = \csc x + \cot x$ Substitute identities for $\csc x$ and $\cot x$.

$\dfrac{\sin x}{1 - \cos x} = \dfrac{1}{\sin x} + \dfrac{\cos x}{\sin x}$ Write as one fraction.

$\dfrac{\sin x}{1 - \cos x} = \dfrac{1 + \cos x}{\sin x}$ Rationalize the numerator.

$\dfrac{\sin x}{1 - \cos x} = \left(\dfrac{1 + \cos x}{\sin x}\right)\left(\dfrac{1 - \cos x}{1 - \cos x}\right)$ Multiply the numerator.

$\dfrac{\sin x}{1 - \cos x} = \dfrac{1 - \cos^2 x}{(\sin x)(1 - \cos x)}$ Replace $1 - \cos^2 x$ with $\sin^2 x$.

$\dfrac{\sin x}{1 - \cos x} = \dfrac{\sin^2 x}{(\sin x)(1 - \cos x)}$ Reduce the common factor of $\sin x$.

$\dfrac{\sin x}{1 - \cos x} = \dfrac{\sin x}{1 - \cos x}$

Most proofs can be done by working right to left or left to right; usually one approach is shorter than the other. Now it's your turn to try a few.

SELF-TEST 2:

Verify the following identities:

1. $\tan x \cot x = 1$
2. $\tan^2 x \sec^2 x - \tan^4 x = \tan^2 x$
3. $\sec x \csc x - \cot x = \tan x$
4. $\csc^2 x(1 - \cos^2 x) = 1$
5. $\cos^3 x \csc^3 x \tan^3 x = \csc^2 x - \cot^2 x$
6. $\dfrac{\cos x + \sin x}{1 + \tan x} = \cos x$
7. $\dfrac{\sec x - \cos x}{\tan x} = \sin x$
8. $2 + \cos^2 x - 3\cos^4 x = (\sin^2 x)(2 + 3\cos^2 x)$
9. $\dfrac{\sec x}{\cos x} - \dfrac{\tan x}{\cot x} = 1$
10. $\sec^2 x \csc^2 x = \sec^2 x + \csc^2 x$

ANSWERS:

1. $\tan x \cot x = 1$
    $\tan x \cdot \dfrac{1}{\tan x} = 1$
    $1 = 1$

2. $\tan^2 x \sec^2 x - \tan^4 x = \tan^2 x$
    $\tan^2 x(\sec^2 x - \tan^2 x) = \tan^2 x$
    $(\tan^2 x)(1) = \tan^2 x$
    $\tan^2 x = \tan^2 x$

3. $\sec x \csc x - \cot x = \tan x$
    $\dfrac{1}{\cos x} \cdot \dfrac{1}{\sin x} - \dfrac{\cos x}{\sin x} = \tan x$
    $\dfrac{1 - \cos^2 x}{\cos x \sin x} = \tan x$
    $\dfrac{\sin^2 x}{\cos x \sin x} = \tan x$
    $\dfrac{\sin x}{\cos x} = \tan x$
    $\tan x = \tan x$

4. $\csc^2 x(1 - \cos^2 x) = 1$
    $\csc^2 x \sin^2 x = 1$
    $\dfrac{1}{\sin^2 x} \cdot \dfrac{\sin^2 x}{1} = 1$
    $1 = 1$

5. $\cos^3 x \csc^3 x \tan^3 x = \csc^2 x - \cot^2 x$
    $\dfrac{\cos^3 x}{1} \cdot \dfrac{1}{\sin^3 x} \cdot \dfrac{\sin^3 x}{\cos^3 x} = \csc^2 x - \cot^2 x$
    $1 = \csc^2 x - \cot^2 x$
    $\csc^2 x - \cot^2 x = \csc^2 x - \cot^2 x$

6. $\dfrac{\cos x + \sin x}{1 + \tan x} = \cos x$
    $\dfrac{\cos x + \sin x}{1 + \dfrac{\sin x}{\cos x}} = \cos x$
    $\dfrac{\cos x + \sin x}{\dfrac{\cos x + \sin x}{\cos x}} = \cos x$
    $\left(\dfrac{\cos x + \sin x}{1}\right)\left(\dfrac{\cos x}{\cos x + \sin x}\right) = \cos x$
    $\cos x = \cos x$

7. $\dfrac{\sec x - \cos x}{\tan x} = \sin x$

$\dfrac{\dfrac{1}{\cos x} - \cos x}{\dfrac{\sin x}{\cos x}} = \sin x$

$\dfrac{\dfrac{1 - \cos^2 x}{\cos x}}{\dfrac{\sin x}{\cos x}} = \sin x$

$\dfrac{\sin^2 x}{\cos x} \cdot \dfrac{\cos x}{\sin x} = \sin x$

$\sin x = \sin x$

8. $2 + \cos^2 x - 3\cos^4 x = (\sin^2 x)(2 + 3\cos^2 x)$
$(2 + 3\cos^2 x)(1 - \cos^2 x) = (\sin^2 x)(2 + 3\cos^2 x)$
$(2 + 3\cos^2 x)(\sin^2 x) = (\sin^2 x)(2 + 3\cos^2 x)$

9. $\dfrac{\sec x}{\cos x} - \dfrac{\tan x}{\cot x} = 1$

$\dfrac{\dfrac{1}{\cos x}}{\cos x} - \dfrac{\dfrac{\sin x}{\cos x}}{\dfrac{\cos x}{\sin x}} = 1$

$\dfrac{1}{\cos x} \cdot \dfrac{1}{\cos x} - \dfrac{\sin x}{\cos x} \cdot \dfrac{\sin x}{\cos x} = 1$

$\dfrac{1}{\cos^2 x} - \dfrac{\sin^2 x}{\cos^2 x} = 1$

$\dfrac{1 - \sin^2 x}{\cos^2 x} = 1$

$\dfrac{\cos^2 x}{\cos^2 x} = 1$

$1 = 1$

10. $\sec^2 x \csc^2 x = \sec^2 x + \csc^2 x$
$(1 + \tan^2 x)(\csc^2 x) = \sec^2 x + \csc^2 x$
$(\csc^2 x + \csc^2 x \tan^2 x) = \sec^2 x + \csc^2 x$
$\csc^2 x + \left(\dfrac{1}{\sin^2 x}\right)\left(\dfrac{\sin^2 x}{\cos^2 x}\right) = \sec^2 x + \csc^2 x$

$\csc^2 x + \dfrac{1}{\cos^2 x} = \sec^2 x + \csc^2 x$

$\csc^2 x + \sec^2 x = \sec^2 x + \csc^2 x$

3 Solving Trigonometric Equations

In this section we'll use our trig identities to solve trigonometric equations. You already know how to solve various types of algebraic equations and how to use the basic trig identities to simplify expressions, so you're off to a good start. As with any equation, the way we solve it is to manipulate the equation into a form where the *variable is isolated*. The following examples will show you different approaches for manipulating the equation into a form that will allow us to solve for the variable.

Example 9:

Solve $\sin^2 x = \dfrac{1}{2}$.

Solution:

To find the value(s) for x that make this statement true we'll rewrite the equation as $(\sin x)^2 = \dfrac{1}{2}$. To solve for x we'll take the positive and the negative square roots of both sides of the equation.

$\sin x = \pm\sqrt{\dfrac{1}{2}}$ Next we'll rationalize the radical by multiplying the numerator

$\sin x = \pm\dfrac{\sqrt{2}}{2}$ and the denominator by $\sqrt{2}$. Now we have to ask ourselves what angle(s) have a sin of $\pm\dfrac{\sqrt{2}}{2}$. In quadrant I the angle is 45°, or $\dfrac{\pi}{4}$. The sign is negative or positive in all quadrants, so the three remaining angles are 135°, 225°, and 315°.

$x = 45°$ or $\dfrac{\pi}{4}$, 135° or $\dfrac{3\pi}{4}$, 225° or $\dfrac{5\pi}{4}$, and 315° or $\dfrac{7\pi}{4}$.

Example 10:

Solve $3 \tan x - 3 = 0$.

Solution:

We'll start by moving the 3 to the other side of the equation.

$3 \tan x = 3$ Now divide both sides of the equation by 3.

$\tan x = 1$ The tangent equals 1 at 45°, or $\dfrac{\pi}{4}$. The tangent is positive in quadrants I and III, so the other angle is 225°, or $\dfrac{5\pi}{4}$.

Analytic Trigonometry

Example 11:
Solve $2\cos^2 x - \cos x = 0$.

Solution:
We'll begin by factoring out the common factor of cos x.
$2\cos^2 x - \cos x = 0$
$\cos x(2\cos x - 1) = 0$ Set each factor equal to 0.

$\cos x = 0 \quad 2\cos x - 1 = 0$ Which angle(s) have a cos of 0? 90° or $\frac{\pi}{2}$, and 270° or $\frac{3\pi}{2}$.

$\cos x = \frac{1}{2}$ Which angle(s) have a cos of $\frac{1}{2}$? The cos is positive in quadrants I and IV. The reference angle that has a cos of $\frac{1}{2}$ is 60°, or $\frac{\pi}{3}$. The angle in quadrant IV that has a cos of $\frac{1}{2}$ is 300°, or $\frac{5\pi}{3}$.

Example 12:
Solve $4\sin x + \csc x = 0$.

Solution:
Sin x and csc x aren't like terms, so we can't combine them. They don't have a common factor, so factoring can't help us. Let's try substituting $\frac{1}{\sin x}$ for csc x to see if that helps us.

$4\sin x + \csc x = 0$

$4\sin x + \dfrac{1}{\sin x} = 0$ We'll write this equation with a common denominator.

$\dfrac{4\sin^2 x + 1}{\sin x} = 0$ The only way a fraction can equal 0 is if the numerator equals 0, so we'll set the numerator equal to 0.

$4\sin^2 x + 1 = 0$
$4\sin^2 x = -1$ Divide both sides by 4.

$$\sin^2 x = -\frac{1}{4}$$

no solution

Normally we would take the square root of both sides of the equation to solve for x, but we can't take the even root of a negative, so the answer is no solution. There isn't any value that will make this equation a true statement.

Example 13:

Solve $\tan x + 3 \cot x = 4$.

Solution:

This problem will require a little more analysis than the others. There aren't any like terms to combine or common factors we can factor out. The only option we see is to substitute $\dfrac{1}{\tan x}$ for $\cot x$. Let's try it to see where that leads us.

$\tan x + 3 \cot x = 4$

$\tan x + 3\left(\dfrac{1}{\tan x}\right) = 4$ Set the equation equal to 0.

$\tan x + \dfrac{3}{\tan x} - 4 = 0$ Write the equation with a common denominator.

$\dfrac{\tan^2 x - 4\tan x + 3}{\tan x} = 0$ Set the numerator equal to 0 and factor.

$(\tan x - 3)(\tan x - 1) = 0$ Set the factors equal to 0 and solve.

$\tan x - 3 = 0,\ \tan x - 1 = 0$

$\tan x = 3,\ \tan x = 1$ The tan is positive in quadrants I and III. We're looking for angles, so we'll use inverse functions.

$x = \tan^{-1} 3,\ x = \tan^{-1} 1$
$x \approx 72°,\ x = 45$

Use your calculator to find $\tan^{-1} 3$. 72° and 45° are in quadrant I; to find the angles in quadrant III we add 180: $72° + 180° = 252°$, and $45° + 180° = 225°$.

Analytic Trigonometry 163

SELF-TEST 3: Solve the given equations for values of x between 0 and 2π:

1. $2\cos x - 1 = 0$
2. $\tan x = -1$
3. $\sqrt{3}\csc x - 2 = 0$
4. $\tan^2 3x = 3$
5. $2\sin x - \tan x = 0$
6. $\tan^2 x - 5\tan x - 6 = 0$
7. $2(2 + \cos x) = 3 + \cos x$
8. $\tan^2 x - 2\sec^2 x + 4 = 0$

ANSWERS:

1. $2\cos x - 1 = 0$
$\cos x = \dfrac{1}{2}$

The cos is positive in quadrants I and IV. In quadrant I, $x = 60°$ or $\dfrac{\pi}{3}$.

In quadrant IV, $x = 300°$ or $\dfrac{5\pi}{3}$.

2. $\tan x = -1$
$x = \tan^{-1}(-1)$

The tangent is negative in quadrants II and IV. We find the reference angle by finding the tangent inverse of -1. This gives us a reference angle of $45°$ or $\dfrac{\pi}{4}$. The corresponding angles in quadrants II and IV are $135°$ or $\dfrac{3\pi}{4}$, and $315°$ or $\dfrac{7\pi}{4}$.

3. $\sqrt{3}\csc x - 2 = 0$

$\csc x = \dfrac{2}{\sqrt{3}}$ Substitute $\dfrac{1}{\sin x}$ for $\csc x$.

$\dfrac{1}{\sin x} = \dfrac{2}{\sqrt{3}}$ Cross-multiply.

$2\sin x = \sqrt{3}$ Divide by 2.

$\sin x = \dfrac{\sqrt{3}}{2}$ The sin is positive in quadrants I and II. The reference angle is $60°$, or $\dfrac{\pi}{3}$. The corresponding angle in quadrant II is $120°$, or $\dfrac{2\pi}{3}$.

4. $\tan^2 3x = 3$
$(\tan 3x)^2 = 3$
$\tan 3x = \pm\sqrt{3}$
$\tan^{-1} \pm \sqrt{3} = 3x$
$3x = 60$
$x = 20$

Take the square root of both sides of the equation. The tangent is positive or negative in all four quadrants. The reference angle whose tangent is $\sqrt{3}$ is 60, or $\dfrac{\pi}{3}$. Remember, we want x, not $3x$, so the value for x in quadrant I is $20°$, or $\dfrac{\pi}{9}$. The corresponding angles in the other quadrants are $160°$, or $\dfrac{8\pi}{9}$; $200°$, or $\dfrac{10\pi}{9}$; and $340°$, or $\dfrac{17\pi}{9}$.

5. $2\sin x - \tan x = 0$ Substitute $\dfrac{\sin x}{\cos x}$ for $\tan x$.

$2\sin x - \dfrac{\sin x}{\cos x} = 0$ Write with a common denominator of $\cos x$.

$\dfrac{2\sin x \cos x - \sin x}{\cos x} = 0$ Set the numerator equal to 0 and factor out a $\sin x$.

$\sin x(2\cos x - 1) = 0$ Set the factors equal to 0 and solve for x.

$\sin x = 0 \quad 2\cos x - 1 = 0$ $\cos x = \dfrac{1}{2}$	The sin of x is equal to 0 at 0° and 180°, or π. The cos of x is positive in quadrants I and IV. The cos equals $\dfrac{1}{2}$ at 60°, or $\dfrac{\pi}{3}$, and at 300°, or $\dfrac{5\pi}{3}$.

6. $\tan^2 x - 5\tan x - 6 = 0$
$(\tan x - 6)(\tan x + 1) = 0$
$\tan x - 6 = 0, \tan x + 1 = 0$
$\tan x = 6, \tan x = -1$

Let's start by factoring this quadratic function.
Set the factors equal to 0 and solve for x.
The tangent is positive in quadrants I and III.
The angle whose tangent is 6 is 81°. Its corresponding angle is 261° in quadrant III. The reference angle whose tangent is 1 is 45°. The tangent is negative in quadrants II and IV. Their corresponding angles are 135° and 315°.

$\tan^{-1} 6 = 81°$

7. $2(2 + \cos x) = 3 + \cos x$
$4 + 2\cos x = 3 + \cos x$
$\cos x = -1$
$x = 180°$ or π

Distribute the 2.
Combine like terms.

8. $\tan^2 x - 2\sec^2 x + 4 = 0$
$\tan^2 x - 2(1 + \tan^2 x) + 4 = 0$
$\tan^2 x - 2 - 2\tan^2 x + 4 = 0$
$2 - \tan^2 x = 0$
$\tan^2 x = 2$
$(\tan x)^2 = 2$
$\tan x = \pm\sqrt{2}$

Substitute $1 + \tan^2 x$ for $\sec^2 x$.
Distribute the −2.
Combine like terms.
Solve for x.

The tangent is positive or negative in all four quadrants. The $\tan^{-1} \sqrt{2} = 55°$ in quadrant I and 125°, 235°, and 305° in the other quadrants.

4 Sum and Difference Formulas

In this section we will use the following formulas to find the exact values of angles that are not standard angles, to simplify trig expressions, and to solve more trig equations.

Table 6.2 Sum and Difference Formulas

$\sin(u + v) = \sin u \cos v + \cos u \sin v \qquad \tan(u + v) = \dfrac{\tan u + \tan v}{1 - \tan u \tan v}$

$\sin(u - v) = \sin u \cos v - \cos u \sin v$

$\cos(u + v) = \cos u \cos v - \sin u \sin v \qquad \tan(u - v) = \dfrac{\tan u - \tan v}{1 + \tan u \tan v}$

$\cos(u - v) = \cos u \cos v + \sin u \sin v$

So far when we've encountered a problem that doesn't have a standard angle, we've used our calculator to estimate the value for x. Sometimes we can find the trig functions of angles that are not standard angles by writing them as a sum or a difference of standard angles. The

Analytic Trigonometry

following examples will show us how to use the sum or the difference formulas to do this.

Example 14:

Find the exact value of sin 75°.

Solution:

Our standard angles in quadrant I are 30°, 45°, and 60°. We can write 75° as 30° + 45°. We can then apply the formula for the sin of the sum of two angles and fill in the values of the trig functions.

sin 75° = sin (30° + 45°) = sin 30° cos 45° + cos 30° sin 45°

$$\left(\frac{1}{2}\right)\left(\frac{\sqrt{2}}{2}\right) + \left(\frac{\sqrt{3}}{2}\right)\left(\frac{\sqrt{2}}{2}\right) = \frac{\sqrt{2}+\sqrt{6}}{4}$$

Before we try a problem involving radians, let's look at different fractional forms of the standard reference angles. This will make it easier for us to rewrite angles as sums or differences of the standard angles.

$$\frac{\pi}{6} = \frac{2\pi}{12} = \frac{3\pi}{18} \qquad \frac{\pi}{4} = \frac{3\pi}{12} \qquad \frac{\pi}{3} = \frac{4\pi}{12} = \frac{6\pi}{18}$$

Example 15:

Find the exact value of $\cos \frac{\pi}{12}$.

Solution:

To figure out how we can rewrite $\frac{\pi}{12}$ as the sum or the difference of two standard angles, look at the reference angles above with a denominator of 12. We could write this problem in either of two ways: $\frac{3\pi}{12} - \frac{2\pi}{12}$ or $\frac{4\pi}{12} - \frac{3\pi}{12}$. We'll show you both ways.

$$\cos \frac{\pi}{12} = \cos\left(\frac{\pi}{4} - \frac{\pi}{6}\right) = \cos \frac{\pi}{4} \cos \frac{\pi}{6} + \sin \frac{\pi}{4} \sin \frac{\pi}{6}$$

$$\left(\frac{\sqrt{2}}{2}\right)\left(\frac{\sqrt{3}}{2}\right) + \left(\frac{\sqrt{2}}{2}\right)\left(\frac{1}{2}\right) = \frac{\sqrt{6}+\sqrt{2}}{4}$$

Now let's use the other method.

$$\cos \frac{\pi}{12} = \cos\left(\frac{\pi}{3} - \frac{\pi}{4}\right) = \cos \frac{\pi}{3} \cos \frac{\pi}{4} + \sin \frac{\pi}{3} \sin \frac{\pi}{4}$$

$$\left(\frac{1}{2}\right)\left(\frac{\sqrt{2}}{2}\right) + \left(\frac{\sqrt{3}}{2}\right)\left(\frac{\sqrt{2}}{2}\right) = \frac{\sqrt{2}+\sqrt{6}}{4}$$

Example 16:

Find the exact value of sin 53° cos 23° − cos 53° sin 23°.

Solution:

By now you should realize that this expression fits into the formula for the sin $(u - v)$, only backward.

sin 53° cos 23° − cos 53° sin 23° = sin (53° − 23°) = sin 30° = $\frac{1}{2}$.

Example 17:

Verify $\tan\left(\frac{\pi}{4} - \theta\right) = \frac{1 - \tan\theta}{1 + \tan\theta}$.

Solution:

We'll use the formula for tangent of a sum.

$$\tan\left(\frac{\pi}{4} - \theta\right) = \frac{1 - \tan\theta}{1 + \tan\theta}$$

$$\frac{\tan\frac{\pi}{4} - \tan\theta}{1 + \tan\frac{\pi}{4}\tan\theta} = \frac{1 - \tan\theta}{1 + \tan\theta}$$

$$\frac{1 - \tan\theta}{1 + \tan\theta} = \frac{1 - \tan\theta}{1 + \tan\theta}$$

Example 18:

Find all solutions of the equation between 0 and 2π.

$$\cos\left(x + \frac{\pi}{4}\right) - \cos\left(x - \frac{\pi}{4}\right) = 1$$

Solution:

We'll start by using the formulas for the cos of a sum and the cos of a difference; then, using the techniques from the previous few sections of this chapter, we'll solve for x.

$$[(\cos x)(\cos \frac{\pi}{4}) - (\sin x)(\sin \frac{\pi}{4})] - [(\sin x)(\sin \frac{\pi}{4}) + (\sin x)(\sin \frac{\pi}{4})] = 1$$

$$(\cos x)\left(\cos \frac{\pi}{4}\right) - (\sin x)\left(\sin \frac{\pi}{4}\right) - (\cos x)\left(\cos \frac{\pi}{4}\right) - (\sin x)\left(\sin \frac{\pi}{4}\right) = 1$$

$$-2 \sin x \sin \frac{\pi}{4} = 1 \qquad\qquad \text{Combine like terms.}$$

$$-(2\sin x)\left(\frac{\sqrt{2}}{2}\right) = 1 \qquad \text{Substitute } \frac{\sqrt{2}}{2} \text{ for } \sin\frac{\pi}{4}.$$

$$-\sqrt{2}\sin x = 1 \qquad \text{Reduce the common factor of 2.}$$

$$\sin x = -\frac{1}{\sqrt{2}} \qquad \text{Rationalize the denominator by multiplying by } \frac{\sqrt{2}}{\sqrt{2}}.$$

$$\sin x = -\frac{\sqrt{2}}{2} \qquad \text{Sin is negative in quadrants III and IV.}$$

$$x = 225° \text{ or } \frac{5\pi}{4}, x = 315°, \text{ or } \frac{7\pi}{4} \qquad \text{The reference angle is } 45°, \text{ or } \frac{\pi}{4}.$$

SELF-TEST 4: Use the trig formulas given in this section to complete the following problems:

1. Find the exact value of the cos 15°. Find it in two ways.

2. Find the exact value of sin 165°.

3. Find the exact value of tan 105°.

4. Find the exact value of $\cos\frac{5\pi}{12}$.

5. Find the exact value of $\frac{\tan 47° - \tan 17°}{1 + \tan 47° \tan 17°}$.

6. Find the exact value of sin 20° cos 110° − cos 20° sin 110°.

7. Solve $\sin\left(x + \frac{\pi}{3}\right) + \sin\left(x - \frac{\pi}{3}\right) = 1$.

8. Verify $\frac{\cos(x-y)}{\cos x \sin y} = \tan x + \cot y$.

9. Verify $\sin(\alpha + \beta) + \sin(\alpha - \beta) = 2\sin\alpha\cos\beta$.

ANSWERS:

1. cos 15°

   First we'll use 45° − 30°, then we'll use 60° − 45°.

   cos 15° = cos (45° − 30°)   Use the formula for the cos of a difference.
   cos 45° cos 30° + sin 45° sin 30°   Substitute values for the angles.

$\left(\dfrac{\sqrt{2}}{2}\right)\left(\dfrac{\sqrt{3}}{2}\right)+\left(\dfrac{\sqrt{2}}{2}\right)\left(\dfrac{1}{2}\right)$ Simplify.

$\dfrac{\sqrt{6}+\sqrt{2}}{4}$

For the second approach we'll use 60° − 45°.

cos 15° = cos (60° − 45°)   Use the formula for the cos of a difference.
cos 60° cos 45° + sin 60° sin 45°   Substitute values for the angles.

$\left(\dfrac{1}{2}\right)\left(\dfrac{\sqrt{2}}{2}\right)+\left(\dfrac{\sqrt{3}}{2}\right)\left(\dfrac{\sqrt{2}}{2}\right)$ Simplify.

$\dfrac{\sqrt{2}+\sqrt{6}}{4}$

2. sin 165° = sin (135° + 30°)   135° is in quadrant II. Its reference angle is 45°.
The sin of 135° is $\dfrac{\sqrt{2}}{2}$, the cos of 135° is $-\dfrac{\sqrt{2}}{2}$.

sin 135° cos 30° + cos 135° sin 30°

$\left(\dfrac{\sqrt{2}}{2}\right)\left(\dfrac{\sqrt{3}}{2}\right)+\left(-\dfrac{\sqrt{2}}{2}\right)\left(\dfrac{1}{2}\right)$ Substitute values for the functions and simplify.

$\dfrac{\sqrt{6}-\sqrt{2}}{4}$

3. tan 105° = tan (45° + 60°)   Use the formula for tangent of a sum.
$\dfrac{\tan 45° + \tan 60°}{1 - \tan 45° \tan 60°}$ Substitute values for the functions and simplify.

$\dfrac{1+\sqrt{3}}{1-(1)(\sqrt{3})}$

$\dfrac{1+\sqrt{3}}{1-\sqrt{3}}$

4. $\cos \dfrac{5\pi}{12} = \cos\left(\dfrac{2\pi}{12}+\dfrac{3\pi}{12}\right)$ Use the formula for the cos of a sum.

$\left(\cos\dfrac{2\pi}{12}\right)\left(\cos\dfrac{3\pi}{12}\right)-\left(\sin\dfrac{2\pi}{12}\right)\left(\sin\dfrac{3\pi}{12}\right)$ Substitute values and simplify.

$\left(\dfrac{\sqrt{3}}{2}\right)\left(\dfrac{\sqrt{2}}{2}\right)-\left(\dfrac{1}{2}\right)\left(\dfrac{\sqrt{2}}{2}\right)$ Simplify.

$\dfrac{\sqrt{6}-\sqrt{2}}{4}$

5. $\dfrac{\tan 47° - \tan 17°}{1 - \tan 47° \tan 17°}$ This fits perfectly into the formula for the tangent of a difference.

$\tan 30° = \dfrac{\sqrt{3}}{3}$

6. sin 20° cos 110° − cos 20° sin 110°   This fits perfectly into the formula for the sin of a difference.

sin (20° − 110°)
sin (−90°) = −1

7. $\sin\left(x+\frac{\pi}{3}\right)+\sin\left(x-\frac{\pi}{3}\right)=1$ Use the formulas for the sin of a sum or a difference.

$(\sin x)\left(\cos\frac{\pi}{3}\right)+(\cos x)\left(\sin\frac{\pi}{3}\right)+(\sin x)\left(\cos\frac{\pi}{3}\right)-(\cos x)\left(\sin\frac{\pi}{3}\right)=1$

$\frac{1}{2}\sin x + \frac{\sqrt{3}}{2}\cos x + \sin x - \frac{\sqrt{3}}{2}\cos x = 1$

$\sin x = 1$

$x = 90°$ or $\frac{\pi}{2}$

8. $\frac{\cos(x-y)}{\cos x \sin y} = \tan x + \cot y$ Use the formula for the cos of a difference.

$\frac{\cos x \cos y + \sin x \sin y}{\cos x \sin y}$ Write as separate fractions.

$\frac{\cos x \cos y}{\cos x \sin y} + \frac{\sin x \sin y}{\cos x \sin y}$ Reduce.

$\frac{\cos y}{\sin y} + \frac{\sin x}{\cos x} = \cot y + \tan x$ Use identities.

9. $\sin(\alpha+\beta) + \sin(\alpha-\beta) = 2\sin\alpha\cos\beta.$ Use the formulas for the sin of a sum or a difference.

$\sin\alpha\sin\beta + \cos\alpha\cos\beta + \sin\alpha\sin\beta - \cos\alpha\cos\beta$ Combine like terms.
$2\sin\alpha\sin\beta.$

5   Multiple-Angle and Product-to-Sum Formulas

In this section we'll show you how to apply more formulas to simplify problems. The more we can simplify a problem, the easier it will be to solve equations. If you can get the hang of simplifying trig equations now, it'll make calculus a lot easier for you.

Double-Angle Formulas

$\sin 2x = 2 \sin x \cos x$ $\tan 2x = \frac{2\tan x}{1-\tan^2 x}$

$\cos 2x = \cos^2 x - \sin^2 x$
$ = 2\cos^2 x - 1$
$ = 1 - 2\sin^2 x$

Example 19:

Use double-angle formulas to verify that the $\cos 60° = \frac{1}{2}$.

Solution:

To use the double-angle formulas we have to write 60° as a double angle, 2(30°). We have three different forms of the cos $2x$ to choose from. We'll use all three to show you how each one works.

cos (60°) = cos 2 (30°) = cos² 30° − sin² 30° = (cos 30°)² − (sin 30°)²

$$= \left(\frac{\sqrt{3}}{2}\right)^2 - \left(\frac{1}{2}\right)^2 = \frac{3}{4} - \frac{1}{4} = \frac{2}{4} = \frac{1}{2}$$

Or cos 2 (30°) = 2 cos² 30° − 1 = 2 (cos 30°)² − 1 = $2\left(\frac{\sqrt{3}}{2}\right)^2 - 1 = 2\left(\frac{3}{4}\right) - 1$

$$= \frac{6}{4} - \frac{4}{4} = \frac{2}{4} = \frac{1}{2}$$

Or cos 2 (30°) = 1 − 2 sin² 30° = 1 − 2(sin 30°)² = $1 - 2\left(\frac{1}{2}\right)^2 = 1 - 2\left(\frac{1}{4}\right)$

$$= 1 - \frac{1}{2} = \frac{1}{2}$$

As you can see, it doesn't matter which form of cos $2x$ we use. We still get $\frac{1}{2}$. As you'll see in the following examples, sometimes one form is easier to use than another.

Example 20:

Use the double-angle formulas to write sin $6x$ in another form.

Solution:

sin $6x$ = sin 2(3x) = 2 sin 3x cos 3x

You may not see the benefit of writing sin $6x$ as 2 sin 3x cos 3x, but it will make solving some equations and performing calculus operations easier. For now you'll just have to trust us. The following example shows how we can use the double-angle formulas in reverse to condense more than one term into one term.

Example 21:

Use a double-angle formula to write cos² $2x$ − sin² $2x$ as a single term.

Solution:

cos² $2x$ − sin² $2x$ = cos 2(2x) = cos 4x

Example 22:

Verify cos $2x$ = 2 cos² x − 1.

Solution:

Our first step is to decide which one of the three versions of cos 2x to start with. We need a $\cos^2 x$ in here, so we need to start with one of the two forms that has a $\cos^2 x$ in it. Now the question is which one we should use. We'll use the one with the $\sin^2 x$ in it because we can't use the one we're trying to prove.

cos 2x
$\cos^2 x - \sin^2 x$ Substitute $1 - \cos^2 x$ for $\sin^2 x$.
$\cos^2 x - (1 - \cos^2 x)$
$\cos^2 x - 1 + \cos^2 x$ Combine like terms.
$2 \cos^2 x - 1$

Example 23:

Solve the trig equation 2 cos x + sin 2x = 0.

Solution:

To solve this equation we have to substitute $2 \sin x \cos x$ for $\sin 2x$.

  2 cos x + sin 2x = 0   Substitute.
2 cos x + 2 sin x cos x = 0   Factor out the cos x.
  cos x (2 + 2 sin x) = 0   Set each factor equal to 0 and solve.
cos x = 0, 2 + 2 sin x = 0

$x = 90°$ or $\dfrac{\pi}{2}$, $x = 270°$ or $\dfrac{3\pi}{2}$

2 + 2 sin x = 0
 2 sin x = –2
  sin x = –1

  $x = 270°$ or $\dfrac{3\pi}{2}$.

Example 24:

Use a double-angle formula to find the tan 120°.

Solution:

We'll use the formula for tan 2x.

$\tan 120° = \tan 2(60°) = \dfrac{2 \tan 60°}{1 - \tan^2 60°} = \dfrac{2\sqrt{3}}{1 - (\sqrt{3})^2} = \dfrac{2\sqrt{3}}{1 - 3} = \dfrac{2\sqrt{3}}{-2} = -\sqrt{3}.$

172 PRECALCULUS

SELF-TEST 5:

1. Rewrite 2 sin 2x using a double-angle formula.
2. Rewrite 6 sin x cos x using a double-angle formula.
3. Rewrite (cos x + sin x)(cos x − sin x) using a double-angle formula.
4. Rewrite $4 - 8 \sin^2 x$ using a double-angle formula.
5. Solve sin 2x − sin x = 0.
6. Solve 4 sin x cos x = 1.
7. Verify $2 + \dfrac{\cos 2x}{\sin^2 x} = \csc^2 x$.
8. Verify $\dfrac{\sin 3x}{\sin x} + \dfrac{\cos 3x}{\cos x} = 4 \cos 2x$.

ANSWERS:

1. 2 sin 2x = 2(2) sin x cos x = 4 sin x cos x

2. 6 sin x cos x
3(2 sin x cos x)
3 sin 2x
    We'll start by writing this in a form that will give us one of the double-angle formulas. Replace 2 sin x cos x by sin 2x.

3. (cos x + sin x)(cos x − sin x)
$\cos^2 x - \sin^2 x = \cos 2x$
    Expand.
Substitute the double-angle formula cos 2x.

4. $4 - 8 \sin^2 x$
$4(1 - 2 \sin^2 x)$
4 cos 2x
    Factor out the common factor of 4.
Substitute the double-angle formula for $1 - 2 \sin^2 x$.

5. sin 2x − sin x = 0
2 sin x cos x − sin x = 0
sin x(2 cos x − 1) = 0
sin x = 0, 2 cos x = 1
$x = 0°, 180°$ or π $\cos x = \dfrac{1}{2}$
$x = 60°$ or $\dfrac{\pi}{3}$, x = 300° or $\dfrac{5\pi}{3}$.
    Use the double-angle formula for sin 2x.
Factor out the sin x.
Set the factors equal to 0 and solve for x.
The cos is positive in quadrants I and IV.

6. 4 sin x cos x = 1
2(2 sin x cos x) = 1
2 sin 2x = 1
$\sin 2x = \dfrac{1}{2}$
$\sin^{-1} \dfrac{1}{2} = 2x$
    Rewrite 4 as 2(2) to create a double-angle formula.
Substitute 2 sin 2x for 2 sin x cos x.
Solve for x.
The sin is positive in quadrants I and II. The angle whose sin is $\dfrac{1}{2}$ is 30°, or $\dfrac{\pi}{6}$, so x = 15° or $\dfrac{\pi}{12}$. In quadrant II it's 165°, or $\dfrac{11\pi}{12}$.

7. $2 + \dfrac{\cos 2x}{\sin^2 x} = \csc^2 x$
$\dfrac{2 \sin^2 x + \cos 2x}{\sin^2 x} = \csc^2 x$
    We'll start by writing the left side of the equation with the common denominator of $\sin^2 x$. Next we'll

Analytic Trigonometry 173

$$\frac{2\sin^2 x + 1 - 2\sin^2 x}{\sin^2 x} = \csc^2 x \qquad \text{substitute } 1 - \sin^2 x \text{ for } \cos 2x.$$

$$\frac{1}{\sin^2 x} = \csc^2 x \qquad \text{Replace } \frac{1}{\sin^2 x} \text{ by } \csc^2 x.$$

$$\csc^2 x = \csc^2 x$$

8.
$$\frac{\sin 3x}{\sin x} + \frac{\cos 3x}{\cos x} = 4\cos 2x \qquad \text{We'll begin by writing this equation with a}$$

$$\frac{\sin 3x \cos x + \cos 3x \sin x}{\sin x \cos x} = 4\cos 2x \qquad \text{common denominator of } \sin x \cos x.$$

$$\frac{\sin(3x+x)}{\frac{1}{2}\sin 2x} \qquad \text{From the last section we see that the numerator is the sin of a sum formula.}$$

$$\frac{2\sin 4x}{\sin 2x} \qquad \text{If we multiply the denominator by 2 and by } \frac{1}{2} \text{ it}$$

would give us $\frac{1}{2}(2\sin x \cos x)$, which

would be $\frac{1}{2}\sin 2x$.

$$\frac{2(2\sin 2x \cos 2x)}{\sin 2x} \qquad \text{Now we'll use the double digit angle formula for the } \sin 4x.$$

$$4\cos 2x = 4\cos 2x \qquad \text{Reduce the common factor of } \sin 2x, \text{ and we're finished.}$$

Don't feel bad if you didn't get this; it's tricky.

Here are a few more formulas for us to try.

Power-Reducing Formulas

$$\sin^2 x = \frac{1 - \cos 2x}{2} \qquad \cos^2 x = \frac{1 + \cos 2x}{2} \qquad \tan^2 x = \frac{1 - \cos 2x}{1 + \cos 2x}$$

Half-Angle Formulas

$$\sin \frac{x}{2} = \pm\sqrt{\frac{1 - \cos x}{2}} \qquad \cos \frac{x}{2} = \pm\sqrt{\frac{1 + \cos x}{2}} \qquad \tan \frac{x}{2} = \frac{1 - \cos x}{\sin x}$$

$$= \frac{\sin x}{1 + \cos x}$$

The signs of $\sin\left(\frac{x}{2}\right)$ and $\cos\left(\frac{x}{2}\right)$ depend on the quadrant in which $\frac{x}{2}$ lies.

Example 25:

Use power-reducing formulas to rewrite $\sin^4 x$ as the sum of first powers of cosines.

Solution:

We'll start by rewriting $\sin^4 x$ as $(\sin^2 x)^2$.

$\sin^4 x = (\sin^2 x)^2 = \left(\dfrac{1-\cos 2x}{2}\right)^2$ Expand $\left(\dfrac{1-\cos 2x}{2}\right)^2$.

$\dfrac{1 - 2\cos 2x + \cos^2 2x}{4}$ Use the power-reduction formula.

$\dfrac{1}{4}\left(1 - 2\cos 2x + \dfrac{1 + \cos 4x}{2}\right)$ Distribute the $\dfrac{1}{4}$.

$\dfrac{1}{4} - \dfrac{1}{2}\cos 2x + \dfrac{1}{8} + \dfrac{\cos 4x}{8}$ Combine like terms.

$\dfrac{3}{8} - \dfrac{1}{2}\cos 2x + \dfrac{1}{8}\cos 4x$ Factor out the $\dfrac{1}{8}$.

$\dfrac{1}{8}(3 - 4\cos 2x + \cos 4x)$

Example 26:

Find the exact value of $\sin 250°$ using a half-angle formula.

Solution:

$\sin 250° = \sin \dfrac{500°}{2} = \sqrt{\dfrac{1 - \cos 500°}{2}}$ Write the $\sin 250°$ as $\dfrac{500°}{2}$.

$\sqrt{\dfrac{1 - \tfrac{1}{2}}{2}} = \sqrt{\dfrac{\tfrac{1}{2}}{2}} = \sqrt{4} = 2 = -2$ The sin is negative in quadrant III.

Here're some more formulas for us to play with:

Table 6.5 Product-to-Sum, Sum-to-Product Formulas

Product-to-Sum Formulas	Sum-to-Product Formulas
$\sin x \sin y = \frac{1}{2}[\cos(x-y) - \cos(x+y)]$	$\sin x + \sin y = 2 \sin\left(\frac{x+y}{2}\right) \cos\left(\frac{x-y}{2}\right)$
$\cos x \cos y = \frac{1}{2}[\cos(x-y) + \cos(x+y)]$	$\sin x - \sin y = 2 \cos\left(\frac{x+y}{2}\right) \sin\left(\frac{x-y}{2}\right)$
$\sin x \cos y = \frac{1}{2}[\sin(x+y) + \sin(x-y)]$	$\cos x + \cos y = 2 \cos\left(\frac{x+y}{2}\right) \cos\left(\frac{x-y}{2}\right)$
$\cos x \sin y = \frac{1}{2}[\sin(x+y) - \sin(x-y)]$	$\cos x - \cos y = -2 \sin\left(\frac{x+y}{2}\right) \sin\left(\frac{x-y}{2}\right)$

Example 27:
Evaluate $\cos 165° \sin 75°$ exactly using a product-to-sum formula.

Solution:
Let's start by substituting into the $\cos x \sin y$ formula $\cos 165° \sin 75°$.

$\frac{1}{2}[\sin(165° + 75°) - \sin(165° - 75°)]$

$\frac{1}{2}[\sin(240°) - \sin(90°)]$ Fill in the values for sin 240° and sin 90°.

$\frac{1}{2}\left[-\frac{\sqrt{3}}{2} - 1\right]$ Write with a common denominator.

$\frac{1}{2}\left[\frac{-\sqrt{3} - 2}{2}\right]$ Distribute the $\frac{1}{2}$.

$\frac{-\sqrt{3} - 2}{4}$

Example 28:
Use a sum-to-product formula to find the exact value of $\cos 105° - \cos 15°$.

Solution:
$\cos 105° - \cos 15°$

$$-2 \sin \frac{105° + 15°}{2} \sin \frac{105° - 15°}{2}$$

$$-2 \sin \frac{120°}{2} \sin \frac{90°}{2}$$

$$-2 \sin 60° \sin 45°$$

$$-2\left(\frac{\sqrt{3}}{2}\right)\left(\frac{\sqrt{2}}{2}\right)$$

$$-\frac{\sqrt{6}}{2}$$

SELF-TEST 6:

For numbers 1 to 3 use half-angle formulas to find the exact value of:

1. sin 165° 2. $\cos \frac{\pi}{8}$ 3. tan 105°

4. Write as a product $\sin(\alpha + \beta) - \sin(\alpha - \beta)$.

5. Verify $\sin\left(\frac{\pi}{6} + x\right) + \sin\left(\frac{\pi}{6} - x\right) = \cos x$.

6. Write as a sum or a difference $4 \sin \frac{\pi}{3} \cos \frac{5\pi}{6}$.

7. Verify $\cos^4 x - \sin^4 x = \cos 2x$.

ANSWERS:

1. $\sin 165° = \sin \frac{330°}{2}$

$\sqrt{\frac{1 - \cos 330}{2}}$ We're asked to use the half-angle formulas, so we'll double 165° and use 330°.

$\sqrt{\dfrac{1 - \dfrac{\sqrt{3}}{2}}{2}}$ Substitute $\frac{\sqrt{3}}{2}$ for cos 330°.

$\sqrt{\dfrac{\dfrac{2 - \sqrt{3}}{2}}{2}}$ Write with a common denominator.

$\frac{1}{2}\sqrt{2 - \sqrt{3}}$ Simplify the radical $\sqrt{\frac{2 - \sqrt{3}}{4}}$.

2. $\cos \frac{\pi}{8} = \cos \frac{\frac{\pi}{4}}{2}$ We're asked to use the half-angle formulas, so we'll use $\frac{\pi}{4}$.

$\sqrt{\dfrac{1 + \cos \dfrac{\pi}{4}}{2}}$ Write with a common denominator and simplify the radical.

$\sqrt{\dfrac{1 + \dfrac{\sqrt{2}}{2}}{2}}$

Analytic Trigonometry

$$\sqrt{\frac{2+\sqrt{2}}{4}}$$

$$\frac{1}{2}\sqrt{2+\sqrt{2}}$$

3. $\tan 105° = \tan\dfrac{210°}{2}$ We're asked to use the half-angle formulas, so we'll use 210°.

$\dfrac{1-\cos 210°}{\sin 210°}$ or $\dfrac{\sin 210°}{1+\cos 210°}$ We have a choice of two formulas to use for tan.

$\dfrac{1-\left(-\frac{\sqrt{3}}{2}\right)}{-\frac{1}{2}}$ or $\dfrac{-\frac{1}{2}}{1+-\frac{\sqrt{3}}{2}}$ We've worked this problem out both ways.

$\dfrac{1+\frac{\sqrt{3}}{2}}{-\frac{1}{2}}$ or $\dfrac{-\frac{1}{2}}{1+-\frac{\sqrt{3}}{2}}$ Substitute for the sin and cos of 210°.

$\dfrac{2+\sqrt{3}}{2} \cdot -\dfrac{2}{1}$ or $-\dfrac{1}{2} \cdot \dfrac{2}{2-\sqrt{3}}$ Simplify the fractions.

$-2-\sqrt{3}$ or $-\dfrac{1}{2-\sqrt{3}} = -2-\sqrt{3}$

4. $\sin(\alpha+\beta) - \sin(\alpha-\beta)$
$\sin\alpha\cos\beta + \cos\alpha\sin\beta - [\sin\alpha\cos\beta - \cos\alpha\sin\beta]$
$\sin\alpha\cos\beta + \cos\alpha\sin\beta - \sin\alpha\cos\beta + \cos\alpha\sin\beta$
$2\cos\alpha\sin\beta$

5. $\sin\left(\dfrac{\pi}{6}+x\right) + \sin\left(\dfrac{\pi}{6}-x\right) = \cos x$

$\sin\dfrac{\pi}{6}\cos x + \cos\dfrac{\pi}{6}\sin x + \sin\dfrac{\pi}{6}\cos x - \cos\dfrac{\pi}{6}\sin x = \cos x$

$2\sin\dfrac{\pi}{6}\cos x = \cos x$ Substitute the values for sin and cos of $\dfrac{\pi}{6}$.

$2\left(\dfrac{1}{2}\right)\cos x = \cos x$

$\cos x = \cos x$

6. $4\sin\dfrac{\pi}{3}\cos\dfrac{5\pi}{6}$ Use the product-to-sum formula.

$4 \cdot \dfrac{1}{2}\left[\sin\left(\dfrac{\pi}{3}+\dfrac{5\pi}{6}\right) + \sin\left(\dfrac{\pi}{3}-\dfrac{5\pi}{6}\right)\right]$ Add the angles.

$2\left[\sin\dfrac{7\pi}{6} + \sin -\dfrac{\pi}{2}\right]$ Substitute the values for the sin of the angles.

7. $\cos^4 x - \sin^4 x = \cos 2x$ Factor.
$(\cos^2 x + \sin^2 x)(\cos^2 x - \sin^2 x) = \cos 2x$ Substitute 1 for $\cos^2 x + \sin^2 x$.
               $1(\cos 2x) = \cos 2x$ Substitute $\cos 2x$ for $\cos^2 x - \sin^2 x$.

7 Additional Topics in Trigonometry

In the previous couple of chapters you learned how to use the Pythagorean theorem and ratios of the sides of right triangles to solve a triangle (i.e., find all three sides and all three angles). The Pythagorean theorem and the standard ratios of sin, cos, tan, and their reciprocals can be applied only to right triangles. But not all triangles are right triangles. In this chapter we'll:

- use the laws of sines and cosines to solve oblique triangles (which we'll explain below)
- learn how to find the area of various types of triangles
- learn how to solve systems of inequalities to solve linear programming problems
- learn how to write a rational expression as the sum or the difference of rational expressions

1 Law of Sines

In this section and the next we'll show you how to solve oblique triangles. Oblique triangles are triangles that do not have a right angle. Other kinds of angles are acute angles, which are less than 90°, and obtuse angles, which are greater than 90° and less than or equal to 180°. Oblique triangles can consist of three acute angles, or two acute angles and one obtuse angle. For us to solve an oblique triangle, we need certain pieces of information about the triangle. There are four possible combinations that allow us to solve oblique triangles:

- one side and two angles
- two sides and the angle opposite one of them
- two sides and the angle included between the sides
- three sides

The fastest, easiest way to solve an oblique triangle is to use the law of sines. The law of sines is based on the assumption that the *sides of a triangle* are proportional to the *sines of the opposite angles*. The figure below shows two oblique triangles, one with an acute angle A, and the other with an obtuse angle A. The law of sines, which is stated below, uses the ratio of the sides of a triangle to the sine of its corresponding angle.

Law of Sines

$$\frac{a}{\sin A} = \frac{b}{\sin B} = \frac{c}{\sin C}$$

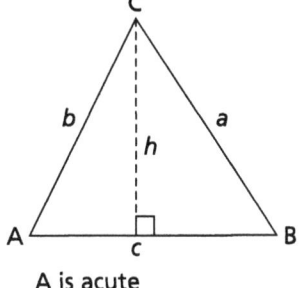
A is acute

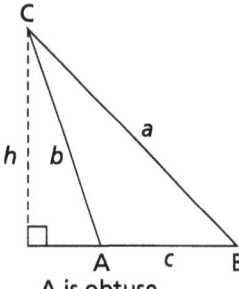
A is obtuse

When we use the law of sines we don't use all three fractions. We use whichever two fractions include the variable we're solving for and allow us to use the given information. Let's work on the following example using the law of sines. It's a lot easier to understand once you've tried a few problems.

Example 1:

Given triangle ABC, A = 50°, B = 30°, b = 10, find a.

Solution:

We're given two angles and the side opposite one of the angles. This is the first case. We can use the law of sines. The figure below shows triangle ABC.

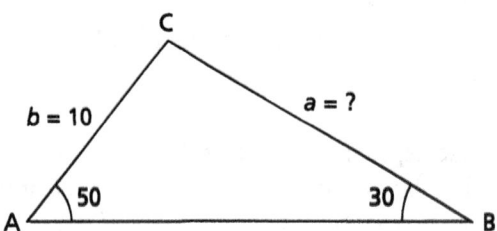

We're looking for a, so we have to use the fraction $\frac{a}{\sin A}$. We know angle B and side b, so the other fraction we will use is $\frac{b}{\sin B}$. Now that we know which fractions to use, let's substitute the values for b, A, and B into $\frac{a}{\sin A} = \frac{b}{\sin B}$.

$\frac{a}{\sin 50°} = \frac{10}{\sin 30°}$ Cross-multiply.

$a \sin 30° = 10 \sin 50°$ Divide both sides by sin 30°.

$a = \frac{10 \sin 50°}{\sin 30°} \approx 15.3209$ We'll round sides to four decimal places.

Now it's time for you to try one.

Example 2:

Solve triangle ABC if c = 6, A = 60°, and B = 40°.

Solution:
The figure below shows triangle ABC.

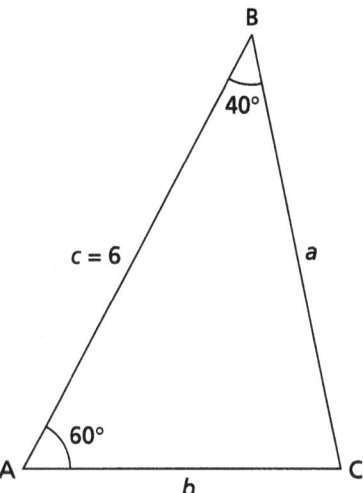

When we're asked to solve a triangle we have to find the measure of all the angles and all the sides. We can easily find the measure of angle C by subtracting the measures of angles A and B from 180°. C = 180° − 60° − 40° = 80°. Now we have to find sides a and b. It doesn't matter which side we find first, so let's start with a. To find a, we have to use the fraction $\dfrac{a}{\sin A}$. We know side c, so the other fraction we'll use is $\dfrac{c}{\sin C}$.

$\dfrac{a}{\sin A} = \dfrac{c}{\sin C}$ Substitute the values for A, C, and c.

$\dfrac{a}{\sin 60°} = \dfrac{6}{\sin 80°}$ Cross-multiply.

$a \sin 80° = 6 \sin 60°$ Divide by sin 80.

$a = \dfrac{6 \sin 60°}{\sin 80°} \approx 5.2763$ Use your calculator to find a.

Now we have to find b:

$\dfrac{b}{\sin 40°} = \dfrac{6}{\sin 80°}$

$b \sin 80° = 6 \sin 40°$

$b = \dfrac{6 \sin 40°}{\sin 80°} \approx 3.9162$

In both examples 1 and 2 we were given two angles and one side. Whenever we're given two angles and one side, one unique triangle is formed from the given information. When we're given two sides and one angle opposite one of the given sides, we have three possibilities:

1. No triangle can be formed from the given information.
2. One unique triangle can be formed from the given information.
3. Two triangles can be formed from the given information.

The table below shows how to figure out how many triangles can be formed from the information given.

Given sides a, b, and angle A or B ($h = b \sin A$), h is the height of the triangle.

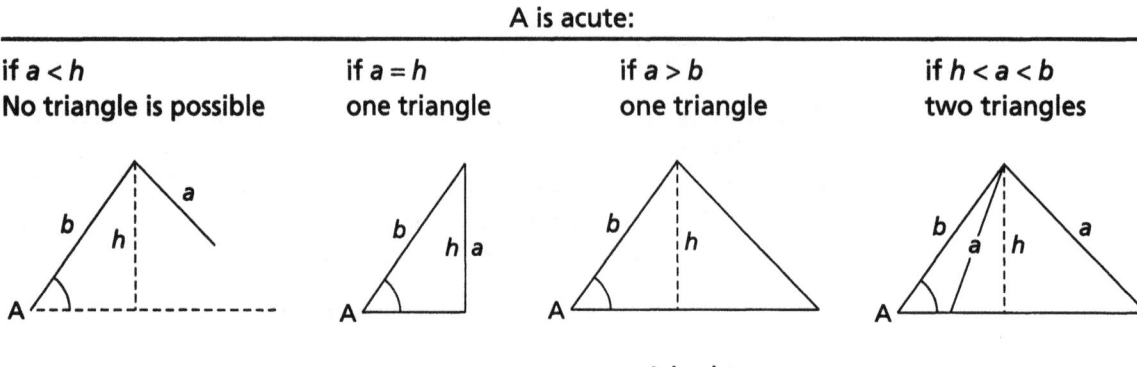

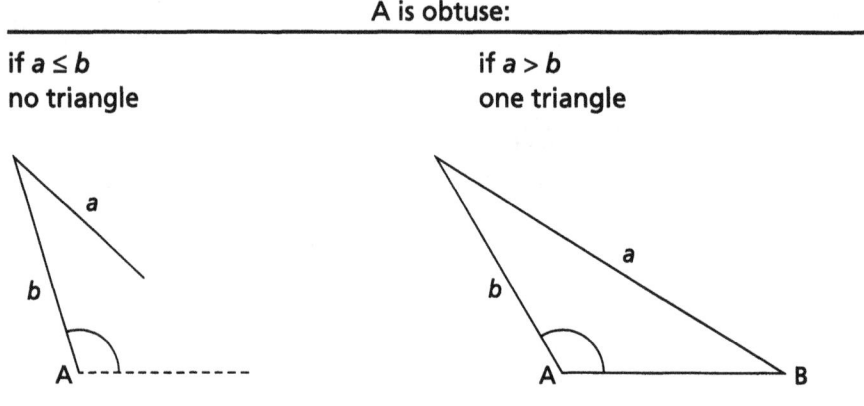

Example 3:

Show that no triangle exists with the given measurements: $a = 15$, $b = 25$, $A = 85$.

Solution:
The figure below shows a drawing of the given information.

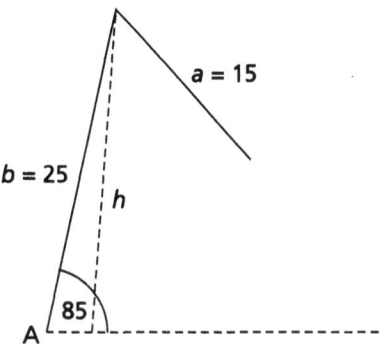

A is acute. $h = b \sin A$ $h = 25 \sin 85 \approx 24.9049$; we see that $a < h$.

If we refer to the table on page 182 we know that no triangle exists with these measurements. If we wanted to verify this we would use $\dfrac{a}{\sin A} = \dfrac{b}{\sin B}$. If we substitute values for the variables we would have

$\dfrac{15}{\sin 85} = \dfrac{25}{\sin B}$ Cross-multiply.

$15 \sin B = 25 \sin 85$ Divide both sides by 15.

$\sin B = \dfrac{25 \sin 85}{15} \approx 1.6603.$ We know that B can't exist because the range of the sin function is between −1 and 1; 1.6603 is not in the range.

Therefore we know that no triangle exists with the given measurements.

Now that you have the hang of the law of sines, it's time to try an example involving two triangles that fit the given measurements.

Example 4:
Solve the triangle with A = 26, a = 1, and b = 1.8.

Solution:
A is an acute angle; $h = b \sin A = 1.8 \sin 26 = .7891$. We know there are two triangles that can be formed with the given measurements because

$h < a < b$. The two triangles are shown below. In the first triangle angle B is acute; in the second, B' is obtuse.

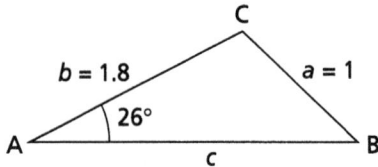

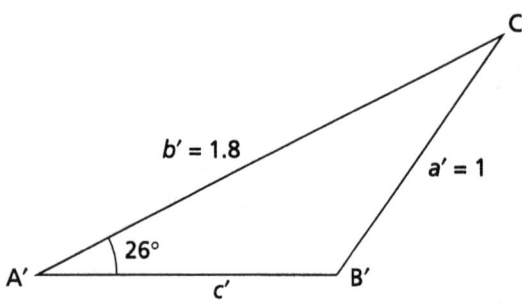

When two triangles are formed, we know that the angles we're looking for would have to be in quadrants I or II. Let's start by finding the measure of angle B.

$$\frac{a}{\sin A} = \frac{b}{\sin B}$$

$$\frac{1}{\sin 26°} = \frac{1.8}{\sin B}$$

$\sin B = 1.8 \sin 26°$

$B = \sin^{-1}(1.8 \sin 26°) \approx 52°$

$B = 180° - 52° = 128°$

$C = 102°, C = 26°$

Once we know the measure of B, we know B' is in quadrant II, so we subtract $180° - 52° = 128°$. We know the sum of the angles of a triangle add to 180°, so we subtract to find C and C'.

Now let's find c:

$$\frac{c}{\sin c} = \frac{a}{\sin A}$$

$$\frac{c}{\sin 102°} = \frac{1}{\sin 26°}$$

$c \sin 26° = \sin 102°$

$$c = \frac{\sin 102°}{\sin 26°} \approx 2.2313$$

Now we use the law of sines to find c and c'.

Now for the other triangle:

$$\frac{c}{\sin C} = \frac{a}{\sin A}$$

$$\frac{c}{\sin 26°} = \frac{1}{\sin 26°}$$

$$c \sin 26° = \sin 26°$$

$$c = \frac{\sin 26°}{\sin 26°} = 1$$

SELF-TEST 1: Use the given information to solve the following triangles using the law of sines:

1. $A = 58°, a = 4.5, b = 12.8$
2. $a = 40, b = 30, A = 75°$
3. $a = 7.07, B = 30°, A = 45°$
4. $A = 58°, a = 4.5, b = 5$
5. $a = 48, A = 110°, b = 16$

ANSWERS:

1. A is an acute angle.
   $h = b; \sin A = 12.8 \sin 58° \approx 10.855$
   $a < h$; therefore no such triangle exists.

2. A is acute.
   $a > b$; therefore one triangle exists with the given measurements.

$\dfrac{a}{\sin A} = \dfrac{b}{\sin B}$	We'll use the law of sines to find the value for B.
$\dfrac{40}{\sin 75°} = \dfrac{30}{\sin B}$	Fill in the values for the variables.
$40 \sin B = 30 \sin 75°$	Cross-multiply.
$\sin B = \dfrac{30 \sin 75°}{40}$	Divide both sides by 40.
$B = \sin^{-1}\left(\dfrac{30 \sin 75°}{40}\right) \approx 46°$	We're looking for an angle, so use $\sin^{-1}$.
$C = 180° - 46° = 134°$	To find C, subtract from 180°.
$\dfrac{c}{\sin C} = \dfrac{a}{\sin A}$	Now we'll use the law of sines to find side c.
$\dfrac{c}{\sin 134°} = \dfrac{40}{\sin 75°}$	Cross-multiply.
$c \sin 75° = 40 \sin 134°$	Divide both sides by sin 75°.
$c = \dfrac{40 \sin 134°}{\sin 75°} \approx 29.7886$	

3. A is acute.

$\dfrac{b}{\sin B} = \dfrac{a}{\sin A}$	We'll use the law of sines to find b.

$$\frac{b}{\sin 30°} = \frac{7.07}{\sin 45°}$$ Cross-multiply.

$b \sin 45° = 7.07 \sin 30°$ Divide both sides by sin 45.

$$b = \frac{7.07 \sin 30°}{\sin 45°} \approx 4.9992$$

$a > b$, so one triangle can be formed from the given measurements.

$$\frac{c}{\sin C} = \frac{a}{\sin A}$$ We'll use the law of sines to find the measure of c.

$$\frac{c}{\sin 105°} = \frac{7.07}{\sin 45°}$$ Cross-multiply.

$c \sin 45° = 7.07 \sin 105°$ Divide both sides by sin 45°.

$$c = \frac{7.07 \sin 105°}{\sin 45°} \approx 9.6578$$

4. A is acute.
$h = b \sin A = 5 \sin 58 \approx 4.2402$
$h < a < b$, so there are two triangles with the given measurements.

$$\frac{4.5}{\sin 58°} = \frac{5}{\sin B}$$ We'll use the law of sines to find B.

$4.5 \sin B = 5 \sin 58°$ Cross-multiply.

$$\sin B = \frac{5 \sin 58°}{4.5}$$ We're looking for an angle, so we'll use $\sin^{-1}$.

$$B = \sin^{-1}\left(\frac{5 \sin 58°}{4.5}\right) \approx 70$$ We know B is 70° in quadrant I. To find B' in quadrant II we subtract: 180° − 70° = 110°.

$B' = 110°$
$C = 180° − 70° − 58° = 52°$
$C' = 180° − 110° − 58° = 12°$ Next, subtract from 180° to find C and C'.

$$\frac{c'}{\sin 12°} = \frac{4.5}{\sin 58°}$$ We'll use the law of sines to find c and c'.

$c' \sin 58° = 4.5 \sin 12°$ Cross-multiply.

$$c' = \frac{4.5 \sin 12°}{\sin 58°} \approx 1.1032$$ Divide both sides by sin 58°.

$$\frac{c}{\sin 52°} = \frac{4.5}{\sin 58°}$$

$c \sin 58° = 4.5 \sin 52°$

$$c = \frac{4.5 \sin 52°}{\sin 58°} \approx 4.1814$$

5. A is obtuse.
$a > b$, so one triangle exists with the given measurements.

$$\frac{48}{\sin 110°} = \frac{16}{\sin B}$$ We'll use the law of sines to find B.

$48 \sin B = 16 \sin 110°$ Use the $\sin^{-1}$ to find the angle.

$$B = \sin^{-1}\left(\frac{16 \sin 110°}{48}\right) \approx 18$$

$C = 180° − 110° − 18° = 52°$.

$$\frac{c}{\sin 52°} = \frac{48}{\sin 110°}$$

$c \sin 110° = 48 \sin 52°$

$$c = \frac{48 \sin 52°}{\sin 110°} \approx 40.252°$$

2 Law of Cosines

If we're given the measure of three sides of a triangle, or two sides and their included angle, we can't use the law of sines. But we can still solve the triangle—using the law of cosines. Listed below are the formulas for the law of cosines. The formulas listed on the left are used to find the measure of a side. The formulas on the right are used to find the measure of an angle. In this section we'll learn how to solve triangles using the law of cosines, and we'll even look at some applications.

Law of Cosines

$$a = \sqrt{b^2 + c^2 - 2bc \cos A} \qquad A = \cos^{-1}\left(\frac{b^2 + c^2 - a^2}{2bc}\right)$$

$$b = \sqrt{a^2 + c^2 - 2ac \cos B} \qquad B = \cos^{-1}\left(\frac{a^2 + c^2 - b^2}{2ac}\right)$$

$$c = \sqrt{a^2 + b^2 - 2ab \cos C} \qquad C = \cos^{-1}\left(\frac{a^2 + b^2 - c^2}{2ab}\right)$$

Example 5:

Given triangle ABC where the sides measure $a = 49.33$, $b = 21.61$, and $c = 42.57$, find the measure of the largest angle.

Solution:

Whenever we're asked to solve a triangle, it's always easiest to find the measure of the largest angle first. *The largest angle is always opposite the longest side.* For this triangle, it's angle A. To find the measure of angle A we'll use the formula from the law of cosines that solves for A:

$$A = \cos^{-1}\left(\frac{b^2 + c^2 - a^2}{2bc}\right).$$

Substitute the values for a, b, and c into the formula.

$$A = \cos^{-1}\left(\frac{(21.61)^2 + (42.57)^2 - (49.33)^2}{2(21.61)(42.57)}\right) \approx 95°$$

Example 6:

Solve triangle ABC where $a = 6$, $b = 7.65$, and $C = 54°$.

Solution:

We're given two sides and the angle between them, so we can't use the law of sines. We can't start by finding angles A or B because we have to know the measure of side c to use those formulas. Let's find c first, then A and B. To find c we'll use this formula:

$$c = \sqrt{a^2 + b^2 - 2ab \cos C}$$
Substitute values for a, b, and C.

$$c = \sqrt{(6)^2 + (7.65)^2 - 2(6)(7.65) \cos 54} \approx 6.369$$

Now that we know the value for c, we can use the law of sines or the law of cosines to find the values of A and B. You haven't had much practice using the law of cosines, so we'll use the law of cosines to find A.

$$A = \cos^{-1}\left(\frac{(7.65)^2 + (6.369)^2 - (6)^2}{2(7.65)(6.369)}\right) \approx 50°$$

If we wanted to find A using the law of sines, we would have done the following:

$$\frac{6}{\sin A} = \frac{6.369}{\sin 54°}$$
$$6 \sin 54° = 6.369 \sin A$$
$$\sin A = \frac{6 \sin 54°}{6.369}$$
$$A = \sin^{-1}\left(\frac{6 \sin 54°}{6.369}\right) \approx 50°$$

Now that we know the value of A, we can find B by subtracting:

$$B = 180° - 54° - 50° = 76°$$

Example 7:

Eight equally spaced holes are drilled in a metal plate on the circumference of a circle 20 cm in radius. What is the center-to-center distance between adjacent holes?

Solution:

Sometimes it's best to begin to solve a word problem by drawing a picture of the problem. The figures below will help us to solve this problem.

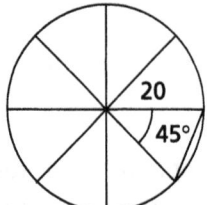
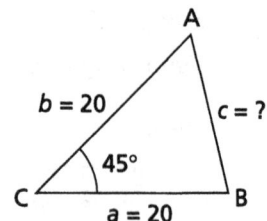

Additional Topics in Trigonometry 189

We know a circle is 360°. This circle is divided into eight equal pieces, so angle C is $\frac{360°}{8} = 45°$. The radius is 20 cm, so we know sides a and b are 20 cm. To find c, we'll substitute these values into the law of cosines formula for c in table 7.2:

$$c = \sqrt{20^2 + 20^2 - 2(20)(20)\cos 45°} = 15.3073 \text{ cm}.$$

Now it's your turn to try a few.

SELF-TEST 2:

Solve the following triangles:

1. $b = 20$, $c = 10$, $A = 60°$
2. $a = 23.31$, $b = 27.26$, $c = 29.17$

3. A room is in the shape of a regular hexagon (six sides). If each side is 10.5 ft, what is the length of the shortest diagonal of the room?

4. Three circles of radius 2, 5, and 8 in are tangent to one another. Find the three angles formed by the lines joining their centers. Round to the nearest angle.

5. Find the perimeter of a regular pentagon inscribed in a circle of radius 12.6 cm.

6. Jean and Fred are standing on level ground and holding ropes attached to a hot-air balloon. Jean's rope is 125 yd long, and Fred's is 105 yd long. If the ropes form an angle of 132.1°, how far apart are Jean and Fred standing?

ANSWERS:

1. We're given the measures of two sides and the included angle, so we have to use the law of cosines. The only formula we can use with the given information is the one to find side a. Once we find the measure of side a, then we can find the measure of angles B or C.

$a = \sqrt{20^2 + 10^2 - 2(20)(10)\cos 60} \approx 17$ $C = \cos^{-1}\left(\frac{17^2 + 20^2 - 10^2}{2(17)(20)}\right) \approx 30°$

$B = 180° - 30° - 60° = 90°$

2. We're given the measures of three sides but no angles, so we have to use the law of cosines. With the given information we can start by using the law of cosines formulas to find any one of the angles. We'll start by finding the measure of angle A, then B, and then subtract from 180° to find C.

$A = \cos^{-1}\left(\frac{(27.26)^2 + (29.17)^2 - (23.31)^2}{2(27.26)(29.17)}\right) \approx 49°$

$B = \cos^{-1}\left(\frac{(23.31)^2 + (29.17)^2 - (27.26)^2}{2(23.31)(29.17)}\right) \approx 61°$

$C = 180° - 49° - 61° = 70°$

3. Remember, a diagonal is a straight line that connects any two nonadjacent vertices. The figures below will help us understand the question better. We have six sides, which means that each of the inside angles formed is 60°. We've drawn a triangle to represent one of the triangles with the shortest diagonal. We know the measure of two sides and the angle between them. The side opposite angle B is the diagonal. To find its measure we have to use the law of cosines formula for side b.

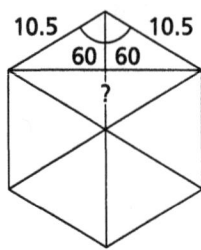
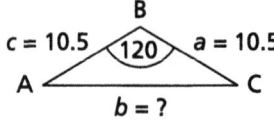

$$b = \sqrt{(10.5)^2 + (10.5)^2 - 2(10.5)(10.5)\cos 120} \approx 18$$

4. The figures below shows the three tangent circles and the triangle formed by the lines connecting their centers. We know the measures of all the sides, but we don't know the measure of any of the angles, so we'll have to use the law of cosines. We can begin by finding the measure of any of the angles. Let's start with angle A. Once we know the measure of angle A, we can use the law of sines to find the measure of one of the other angles. To find the measure of the remaining angle we'll subtract the sum of those two angles from 180.

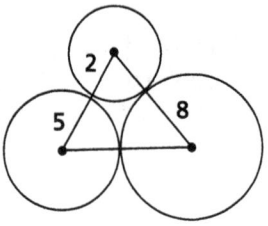
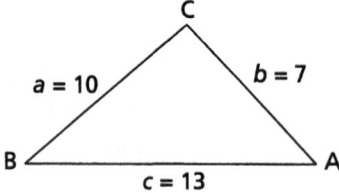

$$A = \cos^{-1}\left(\frac{7^2 + 13^2 - 10^2}{2(7)(13)}\right) \approx 50°$$

$$\frac{b}{\sin B} = \frac{a}{\sin A}$$

$$\frac{7}{\sin B} = \frac{10}{\sin 50°}$$

$$7 \sin 50° = 10 \sin B$$

$$\sin B = \frac{7 \sin 50°}{10}$$

$$B = \sin^{-1}\left(\frac{7 \sin 50°}{10}\right) \approx 32°$$

$$C = 180° - 50° - 32° = 98°$$

5. A pentagon has five sides, so the measure of its inside angles is $\frac{360°}{5} = 72°$. To find its perimeter we have to find the measure of one of its sides and multiply by 5. The following figures show one of the triangles formed inside the pentagon. We know the measure of two of its sides and the included angle. We'll have to use the law of cosines to find the measure of side c.

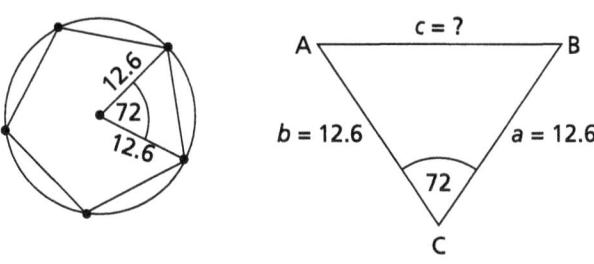

$c = \sqrt{(12.6)^2 + (12.6)^2 - 2(12.6)(12.6)\cos 72°} \approx 15$ $P \approx 15(5) = 75$ cm

6. The following figure illustrates the problem. We know two sides and the included angle, so we have to use the law of cosines.

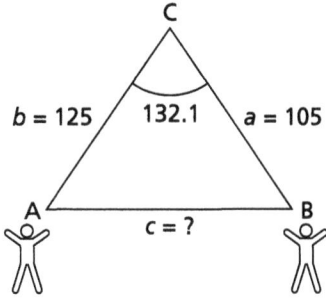

$c = \sqrt{105^2 + 125^2 - 2(105)(125)\cos 132.1°} \approx 210$ yd

3 Area of a Triangle

Sometimes it is necessary to find the area of a triangle, in addition to the angles and sides. The basic formula for the area of a triangle is $A = \frac{1}{2}bh$, where b is the base and h is the height of the triangle. This formula is useful only if we know the base and the height of the triangle. In the following examples we'll work on a couple of problems in which we can use this formula to find the area of the given triangle. Then we'll give you another formula we can use to find the area of an oblique triangle, when we're not given the base and the height of the triangle.

Example 8:

Find the area of a triangular region with a base of 6 and a height of 4.

Solution:

We'll use the formula for the area of a triangle:

$$A = \frac{1}{2}bh = \frac{1}{2}(6)(4) = 12$$

Example 9:

Find the area of the figure below.

Solution:

First we have to find the height, by dropping a perpendicular bisector in one of the triangles. This will create a right triangle, so we can use the tan of half of the 60° angle to find the height of the triangle. Once we have the height of one of the triangles, we can calculate the area of one of the triangles. We can assume the triangles are of equal area because they all have the same measurements and share a side. To find the area of the figure we multiply the area of one triangle by 3.

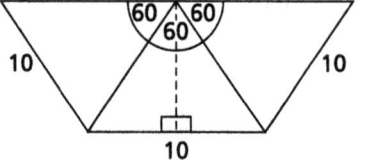

$$\tan \frac{60°}{2} = \frac{5}{h}$$

$$h \tan 30° = 5$$

$$h = \frac{5}{\tan 30°} \approx 8.66$$

$$A = \frac{1}{2}(10)(8.66) \approx 43.3$$

The area of the figure is $A = 3(43.3) = 129.9$.

To find the area of an oblique triangle when we don't know the height or the base, we can multiply the lengths of two sides with the sin of their included angle and divide by 2. This formula is called Heron's formula:

$$A = \frac{1}{2}bc \sin A \text{ or } A = \frac{1}{2}ab \sin C \text{ or } A = \frac{1}{2}ac \sin B$$

Example 10:

Find the area of the triangle below.

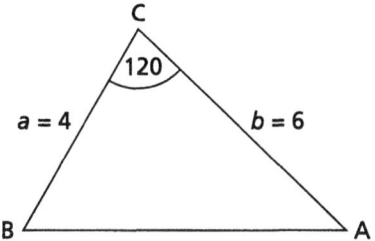

Solution:

We can use Heron's formula.

Area $= \frac{1}{2}ab \sin C = \frac{1}{2}(4)(6) \sin 120 \approx 10.3923$

Example 11:

Find the area of the triangle with vertices (1,0), (2,2), and (4,3).

Solution:

The figure below shows the triangle with the given vertices. We can't use Heron's formula because we don't know the measure of any of the angles. We can't find the measure of any of the angles if we don't know the measure of any of the sides. Our first step is to use the distance formula (from page 24) to find the measure of the sides. After we know the measure of the sides, we can find the measure of one of the angles, and then we can use Heron's formula.

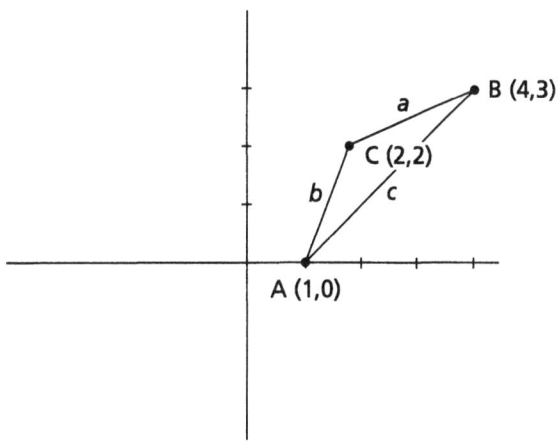

194 PRECALCULUS

$$b = d_{AC} = \sqrt{(2-1)^2 + (2-0)^2} = \sqrt{1+4} = \sqrt{5}$$
$$c = d_{AB} = \sqrt{(4-1)^2 + (3-0)^2} = \sqrt{9+9} = \sqrt{18}$$
$$a = d_{BC} = \sqrt{(4-2)^2 + (3-2)^2} = \sqrt{4+1} = \sqrt{5}$$
$$C = \cos^{-1}\left(\frac{(\sqrt{5})^2 + (\sqrt{5})^2 - \sqrt{18}^2}{2(\sqrt{5})(\sqrt{5})}\right) \approx 143°$$
$$\text{Area} = \frac{1}{2}ab \sin C = \frac{1}{2}(\sqrt{5})(\sqrt{5}) \sin 143° \approx 1.5045$$

SELF-TEST 3:

1. Find the area of the triangle with $B = 72.5°$, $a = 105$, and $c = 64$.
2. Find the area of the triangle with $b = 12.6$, $c = 15$, and $C = 72°$.
3. Find the area of the triangle with vertices (0,2), (–1,–2), and (1, –2).
4. Find the area of the triangle with vertices (–2,1), (1,6), and (3, –1).

ANSWERS:

1. We can use Heron's formula.

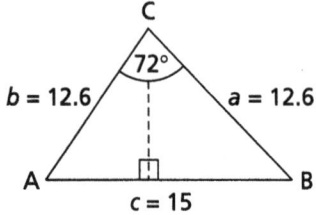

$$\text{Area} = \frac{1}{2} ac \sin B = \frac{1}{2}(105)(64) \sin 72.5° \approx 3{,}204.49$$

2. We can't use Heron's formula because we don't know the measure of A. We'll start by drawing the triangle and dropping a perpendicular bisector. Next we'll solve for h. Once we know h, we don't need to use Heron's formula; we can use the standard formula $A = \frac{1}{2}bh$.

$$\cos 36° = \frac{h}{12.6}$$
$$h = 12.6 \cos 36° \approx 10$$
$$A = \frac{1}{2}bh = \frac{1}{2}(15)(10) = 75$$

3. If we start by plotting the vertices, we can easily see that the base is 2 and the height is 4.

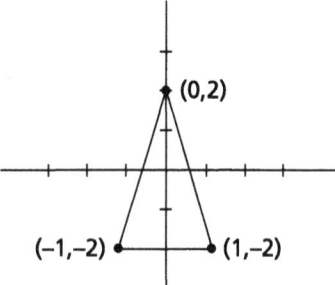

$A = \frac{1}{2}bh = \frac{1}{2}(2)(4) = 4$

4. If we start by plotting the vertices, we can't see the measures of the base or height of this triangle. We'll have to use the distance formula to find the measures of the sides. Then we'll use the law of cosines to solve for A. Once we know the measure of A, we'll use Heron's formula to find the area of the triangle.

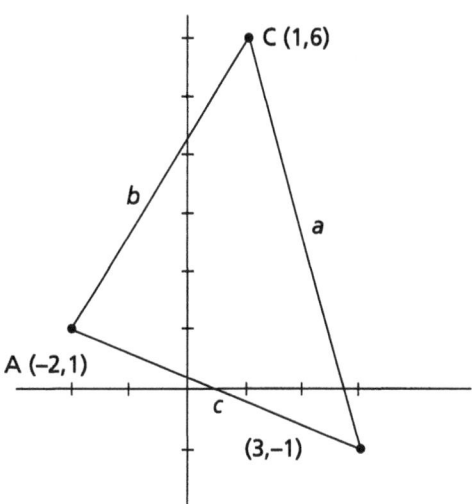

$a = d_{BC} = \sqrt{(3-1)^2 + (-1-6)^2} = \sqrt{4+49} = \sqrt{53}$
$b = d_{AC} = \sqrt{(-2-1)^2 + (1-6)^2} = \sqrt{9+25} = \sqrt{34}$
$c = d_{AB} = \sqrt{(-2-3)^2 + (1--1)^2} = \sqrt{25+4} = \sqrt{29}$
$A = \cos^{-1}\left(\frac{b^2 + c^2 - a^2}{2bc}\right) = \cos^{-1}\left(\frac{34 + 29 - 53}{2(\sqrt{34})(\sqrt{29})}\right) \approx 81$
Area $= \frac{1}{2}bc \sin A = \frac{1}{2}(\sqrt{34})(\sqrt{29}) \sin 81° \approx 15.51$

8 Miscellaneous Topics

1 Solving Systems of Inequalities and Linear Programming

In elementary algebra you learned how to solve systems of equations. The solution to a system of linear equations is the point where the graphs of the lines intersect. This is the only point that's on both lines. In this section we'll show you how to find the solution to a system of linear inequalities. The solution to a system of linear inequalities is every point in a region of the graph where the inequalities overlap, rather than just the point of intersection of the lines. We'll use the following step-by-step procedure to find the solution to a system of linear inequalities.

Steps to Find the Solution to a System of Inequalities

Step 1: Graph the equations of the given lines. If the inequality is < or > use a dotted line. If the inequality is ≤ or ≥ use a solid line. The use of a solid line indicates that the points on the line are part of the solution set.

Step 2: After you have graphed a line, chose any point that is not on the line. We'll call this point a test point. Substitute the test point into the inequality. If it creates a true statement, shade the side of the graph where the test point is located. If it creates a false statement, shade the side of the line opposite where the test point is located. Do this for both lines, one at a time.

Step 3: The solution to the system is indicated by an overlap in the shading. That means that if we substituted any point from the overlapping area into either of the inequalities, it would check.

Let's look at an example. We'll use our step-by-step procedure. You'll have no trouble.

Example 1:

Solve this system:

$2x + 4y < 10$
$3x + 5y \geq 13$

Solution:

Step 1 instructs us to graph the equations, not the inequalities. As we're sure you already know, to graph linear equations, we need two points for each line. You can use any two points on the line. We prefer to use the intercepts whenever possible. In chapter 3 we explained in detail how to find the intercepts of linear equations.

The intercepts of $2x + 4y = 10$ are $\left(0, \frac{5}{2}\right)$, $(5,0)$. The intercepts of $3x + 5y = 13$ are $\left(0, \frac{13}{5}\right)$, $\left(\frac{13}{3}, 0\right)$. The figure below shows the graph of $2x + 4y < 10$.

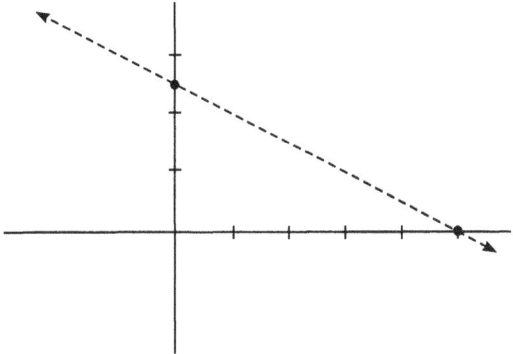

Notice that we used a dotted line because of the inequality (<). Now it's time for step 2, picking a test point. Remember, a test point can be any point that is not on the line. Our favorite test point is (0,0) because it's such an easy substitution. We can use (0,0) for this one because it's not on the line. If we substitute (0,0) into the inequality we have:

2(0) + 4(0) < 10
0 + 0 < 10

0 < 10 is a true statement, so we'll shade the side of the line where (0,0) is.

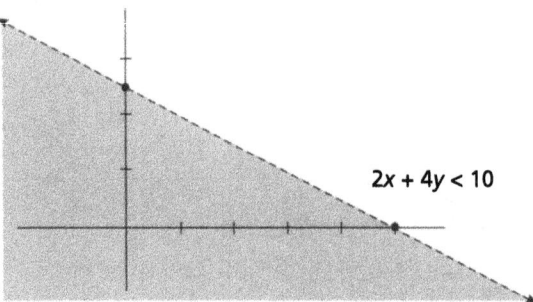

Now we'll draw the graph of $3x + 5y \geq 13$ on the same system. This time we'll use a solid line. We find our intercepts by substituting 0 for x and solving to find its corresponding y value, $\frac{13}{5}$. The y-intercept is $(0, \frac{13}{5})$. Next we'll substitute 0 for y and solve for its corresponding x value, $\frac{13}{3}$. The x-intercept is $(\frac{13}{3}, 0)$. We'll substitute our test point of (0,0) into the inequality.

3(0) + 5(0) ≥ 13
0 + 0 ≥ 13

0 ≥ 13 is a false statement, so we'll shade the side opposite the point (0,0). The section of the graph where the shading overlaps is the solution to the system.

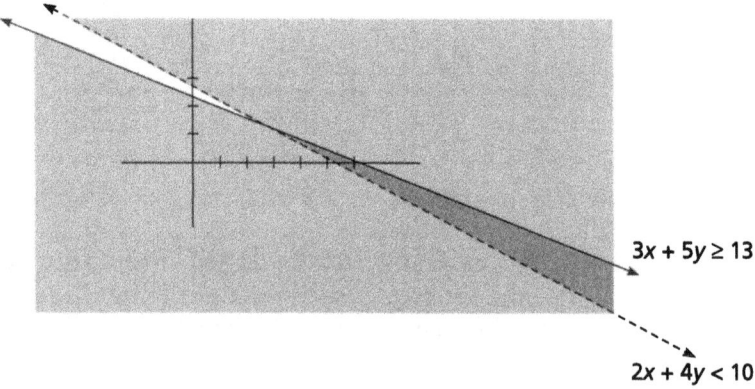

Example 2:

Solve this system:

$$x > 3$$
$$y \leq 4$$
$$3x - 5y \leq 15$$

Solution:

We can start by graphing the first two restrictions. To do that we have to graph the lines $x = 3$ and $y = 4$. The first line is dotted and the second is solid. We'll shade in the area where $x > 3$ and $y < 4$.

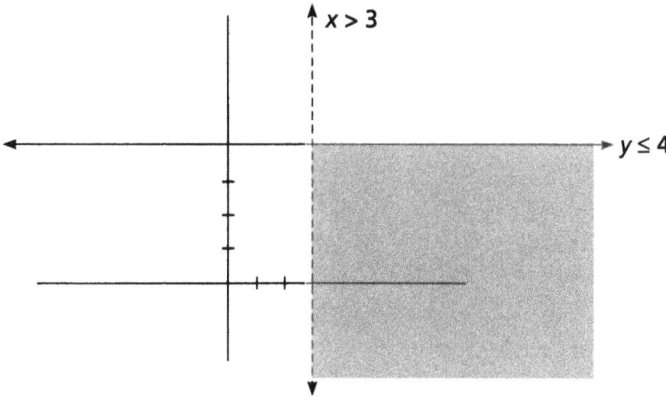

Next we have to graph the line $3x - 5y = 15$. Its intercepts are $(0,-3)$ and $(5,0)$.

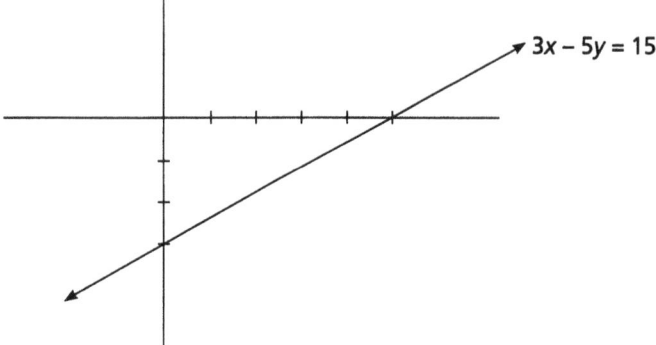

We'll use our favorite test point (0,0) to see which side of the line to shade.

$3(0) - 5(0) \leq 15$
$0 - 0 \leq 15$

$0 \leq 15$ is a true statement, so we'll shade the side of the line where (0,0) is.

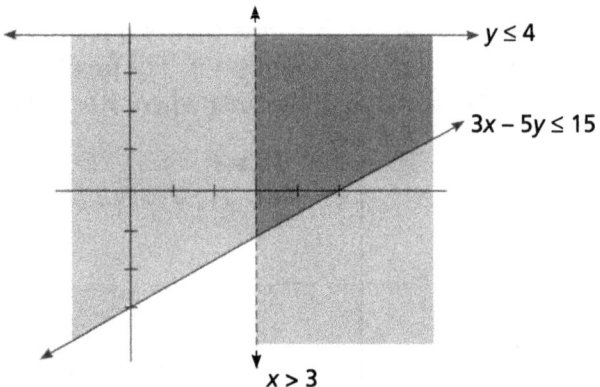

The solution to the system is where all the shaded areas overlap.

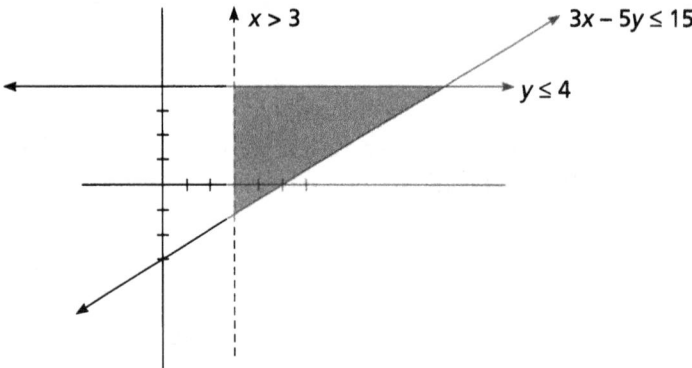

Now it's time to apply what we've just learned to word problems. The type of problem that involves solving systems of inequalities is called linear programming.

To solve a linear programming problem, we have to solve the system of inequalities the same way we did in the previous two examples. We also have to find the points of intersection of the inequalities. We call these points the boundary points. The last thing we have to do is test to see which boundary point yields the maximum or minimum value of the function we're given to work with. Once we work our way through the next problem, it'll be a lot easier for you to understand how to solve a linear programming problem.

Example 3:

A company produces two types of television sets: a basic model and a deluxe model. For each basic model sold there is a profit of $80, and for each deluxe model the profit is $150. The same amount of materials is used to make each model, but the supply is sufficient only for 450 televisions per day. The deluxe model requires twice the time to produce as the regular model. If only regular models were made, there would be time enough to produce 600 per day. Assuming all models will be sold, how many of each model should be produced to get the maximum profit?

Solution:

Let x equal the number of regular televisions produced. Let y equal the number of deluxe televisions produced. The profit equation is $P = 80x + 150y$. Together the number of televisions produced is no more than 450 per day. We'll write that as $x + y \leq 450$. The deluxe model requires twice the time to produce as the regular model. We'll write that as $x = 2y$. If only regular models were made, there would be time enough to produce 600 per day. We'll write that as $x \leq 600$. We're assuming both models will be produced, so we're going to combine those last two statements into one: $x + 2y \leq 600$. The two inequalities we have to graph are $x + y \leq 450$ and $x + 2y \leq 600$. We know the graph has to be restricted to the first quadrant because the other quadrants have negative values. We can't produce a negative number of products. We'll use the same technique we used in the previous two examples to graph the system.

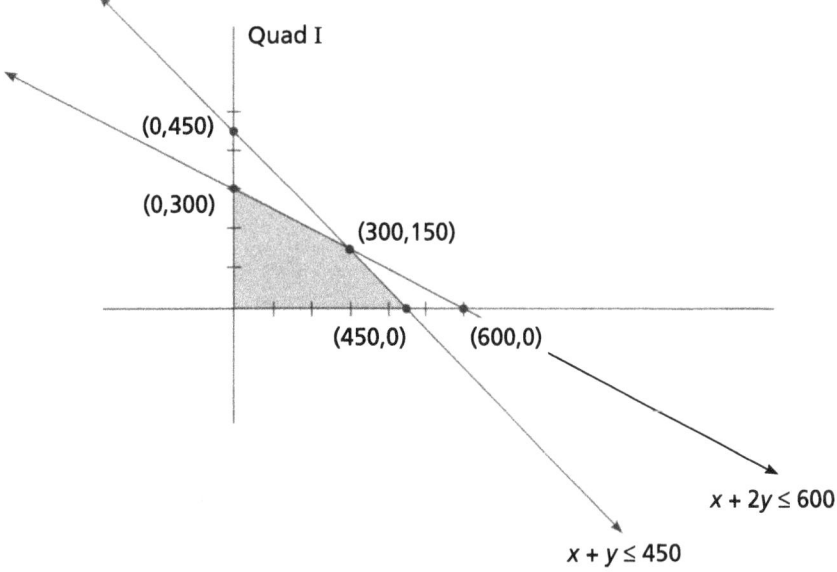

The boundary points are the points that box in the solution to the system. We can see that there are three boundary points. They are (0,300), (450,0), and the last one is the intersection of the two lines. If we solve the system of equations we find that the coordinates of the last boundary point are (300, 150). To see which points will yield the maximum profit we substitute each boundary point into the profit function.

$P = 80x + 150y$
For (0,300) the profit is $P = 80(0) + 150(300) = \$45,000$.
For (450,0) the profit is $P = 80(450) + 150(0) = \$36,000$.
For (300,150) the profit is $P = 80(300) + 150(150) = \$46,500$.

The maximum profit is reached when 300 regular and 150 deluxe televisions are produced.

SELF-TEST 4:

Solve the following systems:

1. $3x + 4y < -12$
   $2x - 3y \leq 6$

2. $\quad x > -1$
   $\quad y \geq 1$
   $x - 2y > 2$
   $x + 3y < 9$

3. Melanie supplies a wholesaler with two mixtures of nuts packed in 8-lb cans. The first mixture contains 6 lb of peanuts and 2 lb of cashews in each can. The second mixture contains 5 lb of peanuts and 3 lb of cashews. The profit on the first mixture is $4 per can, and the profit on the second is $5 per can. From a supply of 240 lb of peanuts and 96 lb of cashews, how many cans of each mixture should be made for a maximum profit?

ANSWERS:

1. Let's start by finding the intercepts of the graphs and sketching the graphs. Then we'll use our favorite test point of (0,0) to see which side of the lines to shade. The solution to the system is the section of the graph where the shading overlaps.

$3x + 4y < -12$ (0,–3) (–4,0) $2x - 3y \leq 6$ (0,–2) (3,0)
test (0,0)   test (0,0)
$3(0) + 4(0) < -12$ $2(0) - 3(0) \leq 6$
$\quad\quad 0 < -12$ $\quad\quad 0 \leq 6$
false statement   true statement
shade the side opposite (0,0)   shade the same side as (0,0)

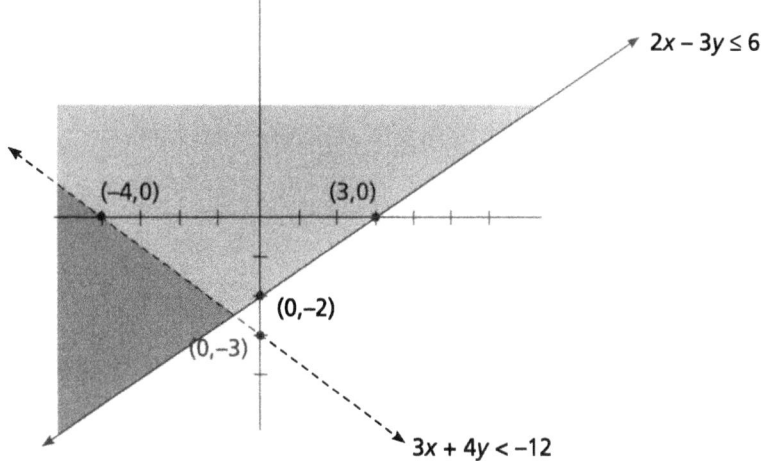

2. All x values are to the right of −1. All y values are above 1.
x − 2y > 2 (0,−1), (2,0)   x + 3y < 9 (0,3), (9,0)
test (0,0)   test (0,0)
0 − 2(0) > 2   0 + 3(0) < 9
    0 > 2   0 < 9
false statement   true statement
shade the side opposite (0,0)   shade the same side as (0,0)

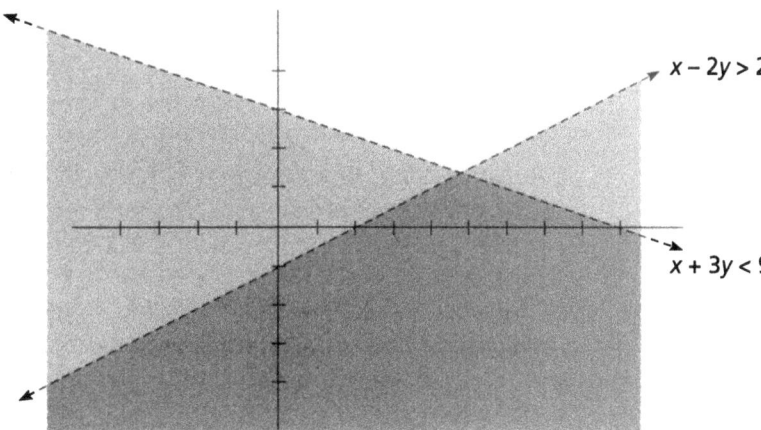

3. Let x = the pounds of peanuts. Let y = the pounds of cashews. The profit is P = 4x + 5y. We're restricted to 240 lb of peanuts and 96 lb of cashews. The equations are:
6x + 5y ≤ 240   2x + 3y ≤ 96
(0,48), (40,0)   (0,32), (48,0)
test (0,0)   test (0,0)
6(0) + 5(0) ≤ 240   5(0) + 3(0) ≤ 96
     0 ≤ 240   0 ≤ 96
true statement   true statement
shade the same side as (0,0)   shade the same side as (0,0)

Solve the system to find the third boundary point. The boundary points are (0,32), (30,12), and (40,0). The profit is $P = 4x + 5y$.
For (0,32) $P = 4(0) + 5(32) = \$160$.
For (30,12) $P = 4(30) + 5(12) = \$180$.
For (40,0) $P = 4(40) + 5(0) = \$160$.

The maximum profit occurs when there are 30 cans of the first mixture and 12 cans of the second mixture.

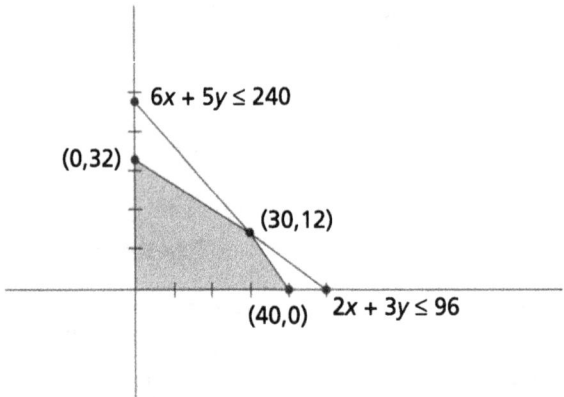

2. Partial Fraction Decomposition

Sometimes, to apply various types of operations to a rational expression, it is necessary to rewrite the rational expression as the sum or the difference of two or more rational expressions. When we rewrite a rational expression as the sum or the difference of two or more rational expressions, it's called partial fraction decomposition. This situation arises fairly often in calculus problems. Before we show you how to write a rational expression as the sum or the difference of rational expressions, let's do the opposite. Let's write a sum or a difference of rational expressions as a single rational expression. Although this is something you already learned in a basic algebra course, it will help you to understand how to perform partial fraction decomposition.

Example 4:

Write $\dfrac{2}{x-7} + \dfrac{3}{x+2}$ as a single fraction.

Solution:

To add two fractions, we find their common denominator. In this case it's $(x-7)(x+2)$. Next we convert both fractions to the common denominator by multiplying the numerators and the denominators by the necessary factors. We combine the terms in the numerators and write them over the common denominator.

$$\frac{2(x+2)+3(x-7)}{(x-7)(x+2)} = \frac{2x+4+3x-21}{(x-7)(x+2)} = \frac{5x-17}{x^2-5x-14}$$

We just wrote the sum of two fractions as one fraction. In example 5 we're going to do the exact opposite. We're going to take the answer from example 4 and write it as the sum of two fractions. This is our first partial fraction decomposition problem.

Example 5:

Write $\frac{5x-17}{x^2-5x-14}$ as the sum of two fractions.

Solution:

The first step on all partial fraction decomposition problems is to check to see if the fraction is improper. Remember, a fraction is improper if the degree of the numerator is not less than the degree of the denominator. If it's improper we divide the numerator by the denominator. This fraction is proper—$5x$ is a lesser degree than x^2—so we don't divide. The next step is to factor the denominator. This denominator factors into $(x-7)(x+2)$. Notice in example 4 that the original denominators became the factors of the sum. We'll use this knowledge to write the given fraction as the sum of two fractions whose denominators are the factors. We know there has to be something in the numerators, so we'll fill them in with the variables A and B. As we work through the problem we'll find the values for A and B.

$\frac{5x-17}{x^2-5x-14} = \frac{A}{x-7} + \frac{B}{x+2}$	Our next step is to multiply every term of the equation by the common denominator.
$5x - 17 = A(x+2) + B(x-7)$	This reduces out the denominators.
$5x - 17 = Ax + 2A + Bx - 7B$	Distribute and combine like terms.
$5x = x(A+B),\ -17 = 2A - 7B$	Factor out the x and set the like terms equal.
$A + B = 5$ and $2A - 7B = -17$	Now we have a system to solve.

$$A + B = 5$$
$$2A - 7B = -17$$
$$-2A - 2B = -10$$

Multiply the first equation by -2.

Combine the equations and solve for B.

$$2A - 7B = -17$$
$$-9B = -27$$
$$B = 3$$

Substitute $B = 3$ into one of the equations to find A.

$$A + B = 5$$
$$A + 3 = 5$$
$$A = 2$$

Now that we know the values for A and B we can write the partial fraction decomposition.

$$\frac{5x - 17}{x^2 - 5x - 14} = \frac{2}{x - 7} + \frac{3}{x + 2}$$

Example 6:

Write the partial fraction decomposition for $\frac{11x + 6}{x^2 + 2x + 1}$.

Solution:

This is not an improper fraction, so we don't have to divide. We'll factor the denominator first into $(x + 1)^2$. This is a linear factor squared. When we have a factor to a power higher than 1, we have to use more than one fraction. One of the fractions should have the factor to the first power, and a second fraction should have the factor squared.

$$\frac{11x + 6}{x^2 + 2x + 1} = \frac{A}{x + 1} + \frac{B}{(x + 1)^2}$$

Multiply the equation by the common denominator.

$$11x + 6 = A(x + 1) + B$$
$$11x + 6 = Ax + A + B$$
$$A = 11, A + B = 6 \text{ so } B = -5$$

Expand.

$A = 11$ because the coefficient of x on both terms of the equation is 11. $A + B =$ the constant 6; therefore we now have the partial fraction decomposition.

$$\frac{11x + 6}{x^2 + 2x + 1} = \frac{11}{x + 1} - \frac{5}{(x + 1)^2}$$

Example 7:

Write the partial fraction decomposition of $\dfrac{x^2 + x + 3}{x^4 + 6x^2 + 9}$.

Solution:

This is a proper fraction, so we don't have to divide. We'll factor the denominator into $(x^2 + 3)^2$. In this case we have quadratic factors. When we have quadratic factors we can't fill the numerators in with just an A or a B because the numerators don't have to be constants—they could be linear. We'll fill the numerators in with $Ax + B$ and $Cx + D$.

$$\dfrac{x^2 + x + 3}{x^4 + 6x^2 + 9} = \dfrac{Ax + B}{x^2 + 3} + \dfrac{Cx + D}{(x^2 + 3)^2} \qquad \text{Multiply by the least common denominator.}$$

$$x^2 + x + 3 = (Ax + B)(x^2 + 3) + Cx + D \qquad \text{Expand.}$$
$$0x^3 + x^2 + x + 3 = Ax^3 + 3Ax + Bx^2 + 3B + Cx + D \qquad \text{Set the like terms equal.}$$

$0x^3 = Ax^3$	$x^2 = Bx^2$	$x = x(3A + C)$	$3 = 3B + D$
$A = 0$	$B = 1$	$1 = 3A + B$	$3 = 3 + D$
		$C = 1$	$D = 0$

The partial fraction decomposition is:

$$\dfrac{x^2 + x + 3}{x^4 + 6x^2 + 9} = \dfrac{1}{x^2 + 3} + \dfrac{x}{(x^2 + 3)^2}$$

Now it's your turn.

SELF-TEST 5:

Write the partial fraction decomposition of the following fractions:

1. $\dfrac{1}{x^2 - 5x + 6}$
2. $\dfrac{5x^2 + 20x + 6}{x^3 + 2x^2 + x}$
3. $\dfrac{x^2 + x + 3}{x^4 + 6x^2 + 9}$
4. $\dfrac{x + 1}{x^3 + x}$
5. $\dfrac{x^3 - x + 3}{x^2 + x - 2}$

ANSWERS:

1. $\dfrac{1}{x^2 - 5x + 6} = \dfrac{A}{x - 3} + \dfrac{B}{x - 2}$ Multiply by the common denominator.

$1 = A(x - 2) + B(x - 3)$ Distribute and equate like terms.
$0x + 1 = Ax - 2A + Bx - 3B$ We'll write $0x$ because there isn't an x to the first-
$0x = x(A + B) \quad 1 = -2A - 3B$ power term. Solve the system for A and B.
$A + B = 0 \qquad -2A - 3B = 1$

$A + B = 0$
$-2A - 3B = 1$
$2A + 2B = 0$ Multiply the first equation by 2.
$-2A - 3B = 1$ Combine equations and solve for B, then solve for A.
$-B = 1$
$B = -1$
$A = 1$

The partial fraction decomposition is $\dfrac{1}{x^2 - 5x + 6} = \dfrac{1}{x-3} - \dfrac{1}{x-2}$.

2. Factor the denominator into $x(x + 1)^2$. These are all linear factors, so use A, B, and C in the numerators. We have to use three fractions, one with a denominator of x, another with $x + 1$, and the last with $(x + 1)^2$.

$$\dfrac{5x^2 + 20x + 6}{x^3 + 2x^2 + x} = \dfrac{5x^2 + 20x + 6}{x(x+1)^2}$$

$$\dfrac{5x^2 + 20x + 6}{x(x+1)^2} = \dfrac{A}{x} + \dfrac{B}{x+1} + \dfrac{C}{(x+1)^2}$$ Multiply by the common denominator.

$5x^2 + 20x + 6 = A(x+1)^2 + Bx(x+1) + Cx$
$5x^2 + 20x + 6 = A(x^2 + 2x + 1) + Bx(x+1) + Cx$
$5x^2 + 20x + 6 = Ax^2 + 2Ax + A + Bx^2 + Bx + Cx$

$5x^2 = x^2(A+B)$ $20x = x(2A + B + C)$ $6 = A$
$5 = A + B$ $20 = 2A + B + C$ $A = 6$
if $A = 6$, $B = -1$ if $A = 6$ and $B = -1$, $C = 9$

The partial fraction decomposition is $\dfrac{5x^2 + 20x + 6}{x^3 + 2x^2 + x} = \dfrac{6}{x} - \dfrac{1}{x-1} + \dfrac{9}{(x+1)^2}$.

3. We'll start by factoring the denominator into $(x^2 + 3)^2$. Both of the factors are quadratic, so we'll use $Ax + B$ as the numerator for the first fraction. We'll use $Cx + D$ as the numerator of the second fraction.

$$\dfrac{x^2 + x + 3}{(x^2 + 3)^2} = \dfrac{Ax + B}{x^2 + 3} + \dfrac{Cx + D}{(x^2 + 3)^2}$$

$x^2 + x + 3 = (Ax + B)(x^2 + 3) + Cx + D$
$0x^3 + x^2 + x + 3 = Ax^3 + 3Ax + Bx^2 + 3B + Cx + D$

$0x^3 = Ax^3$ $x^2 = Bx^2$ $x = x(3A + C)$ $3 = 3B + D$
$A = 0$ $B = 1$ $1 = 3A + C$ if $B = 1$
        if $A = 0$ $D = 0$
        $C = 1$

The partial fraction decomposition is $\dfrac{x^2 + x + 3}{x^4 + 6x^2 + 9} = \dfrac{1}{x^2 + 3} + \dfrac{x}{(x^2 + 3)^2}$.

4. The denominator factors into $x(x^2 + 1)$. We have one linear factor, x, so we'll use A for its numerator. We have one quadratic factor, $x^2 + 1$, so we'll use $Bx + C$ for its numerator.

$$\dfrac{x+1}{x^3 + x} = \dfrac{A}{x} + \dfrac{Bx + C}{x^2 + 1}$$

$x + 1 = A(x^2 + 1) + x(Bx + C)$
$x + 1 = Ax^2 + A + Bx^2 + Cx$
$0x^2 + x + 1 = Ax^2 + A + Bx^2 + Cx$

$0x^2 = x^2(A + B)$ $x = Cx$ $1 = A$
$A + B = 0$ $C = 1$ $A = 1$
$A = 1$, $B = -1$

The partial fraction decomposition is $\dfrac{x+1}{x(x^2+1)} = \dfrac{1}{x} + \dfrac{-x+1}{x^2+1}$.

5. This is an improper fraction, so we have to divide first.

$$\begin{array}{r} x-1 \\ x^2+x-2 \overline{\smash{)}x^3-x+3} \\ \underline{x^3+x^2-2x} \\ -x^2+x+3 \\ \underline{-x^2-x+2} \\ 2x+1 \end{array}$$

The remainder goes over the divisor.

$$\dfrac{x^3-x+3}{x^2+x-2} = x-1+\dfrac{2x+1}{x^2+x-2}$$

We have to find the decomposition of the remainder. Factor the denominator first.

$$\dfrac{2x+1}{(x+2)(x-1)} = \dfrac{A}{x+2}+\dfrac{B}{x-1}$$

Multiply both sides by the L.C.D. $(x+2)(x-1)$.

$$2x+1 = A(x+1)+B(x+2)$$
$$2x+1 = Ax-A+Bx+2B$$
$$2x = x(A+B)$$
$$2 = A+B \quad\quad 1 = -A+2B$$

Solve the system.

$$A+B=2$$
$$-A+2B=1$$
$$3B=3$$
$$B=1; \text{ therefore } A=1.$$

The partial fraction decomposition is $\dfrac{x^3-x+3}{x^2+x-2} = x-1+\dfrac{1}{x+2}+\dfrac{1}{x-1}$.

Index

acute angles, 178–79, 182
addition
    of functions, 41–42
    of polynomials, 11
algebraic statements, 10
amplitude, of graphs, 141
analytic trigonometry. *See* trigonometry
angles
    acute, 178–79, 182
    coterminal, 128–29, 130–31
    degree form, 129–30
    initial and terminal sides, 128
    inverse trigonometric functions, 143–44
    measurement of, 127–31
    obtuse, 179, 182
    positive and negative, 128
    radians, 129–30
    reference, 134–39
    self-tests, 131, 139
    standard unit circle, 127–28, 130
    sum and difference formulas and, 164–65
    trigonometric functions, 134–40
applications, 1
    basic mathematical concepts, 28–32
    exponential and logarithmic functions, 119–26
    pretest, 3
    trigonometric, 146–51
arcsin, 143
area
    basic geometric formulas, 25–27, 191
    of triangles, 25, 27, 178, 191–95
asymptotes, 82–102
    definition of, 82
    exponential functions, 105
    horizontal, 84, 85–86, 88–92
    oblique, 98–102
    self-tests, 87, 93, 100
    vertical, 83, 84–85, 88–92
axes, intercepts and, 53–57
axis of symmetry, 67, 68, 74

base
    of exponent, 5, 6, 9, 106–8, 117
    logarithmic functions, 109
basic geometry, 1, 24–28
    area formulas, 25–27, 191
    distance formula, 24–25, 193
    midpoint formula, 24–25
    perimeter formulas, 25–27, 191
    pretest, 3
    self-test, 27–28

Index

binomials
  division, 13, 14–15
  factoring, 18, 20
  multiplication, 12, 21–22
  sum/difference of two cubes, 23–24
boundary points, 200, 202
brackets, in notation, 38

carbon dating, 103, 124–25
Cartesian coordinate system, 140–41
circles
  circumference and area formulas, 26, 27
  standard unit, 127–28, 130, 134–36
circumference, of circles, 26, 27
coefficients
  exponents and, 6
  horizontal asymptotes and, 85, 88
  logarithmic functions, 112–13
  parabolas and, 67
  in polynomials, 11, 77
combining like terms, in polynomials, 11
commutative property
  composite functions and, 43
  multiplication, 21
composite functions, 33, 43–45, 49–50
compounded interest, 103, 120–22
condensing, logarithmic functions, 113
conjugates, of denominators, 154
constant of proportionality, 123–25
constants
  Euler's, 115
  exponents and, 6
  intercepts and, 65
  parabolas and, 67, 72–74
  in straight-line equations, 60, 61
coordinates
  Cartesian system, 140–41
  intercepts, 53, 59
  slope of line, 57
  vertices of parabolas, 68–69
cosecant
  graphs of functions, 141–42
  identities, 153
  right triangles, 132–33
cosine
  angles, 134–37
  double-angle formulas, 169
  graphs of functions, 140–41
  half-angle formulas, 173
  identities, 153
  inverse functions, 144
  law of cosines, 187–91
  power-reducing formulas, 173
  product-to-sum formulas, 175
  right triangles, 132–33
  sum and difference formulas, 164
  sum-to-product formulas, 175–76
cotangent
  graphs of functions, 141–42
  identities, 153
  right triangles, 132–33
coterminal angles, 128–29, 130–31
cubes, sum/difference of two, 23–24
curves, period of, 141

decay, exponential, 123, 124–25
decimals, rewriting of, 107
degree form, angles, 129–30
delta (Greek letter), in notation, 57
denominators
  asymptotes and, 85–86, 88, 90, 92
  conjugates of, 154
  exponents and, 7, 8–9
  improper fractions, 98, 205
  in polynomials, 10
  rational exponents and, 16
  rationalizing of, 154
dependent variables, 34–35
descending order, of polynomials, 14
diagonals, 190
difference of two squares
  factoring and, 20–21
  intercepts, 54, 55
difference/sum of two cubes, 23–24
distance formula, basic geometry, 24–25, 193
distributing terms, in polynomials, 11–12
division
  of functions, 41–42
  of polynomials, 13–15
division boxes, 14
division property, logarithmic functions, 112, 113
domain of functions, 33, 39, 52
  acceptable members, 37–39

domain of functions *(continued)*
  asymptotes, 83–84, 85
  definition of, 34–35
  exponential functions, 105
  inverse functions, 45–46, 109
  inverse trigonometric, 143
  trigonometric, 140–42
double-angle formulas, trigonometric, 169–73
double negatives, 15

e. See Euler's constant
equations
  exponential, 104, 105–8
  linear, 62–67, 196–204
  logarithmic, 114–19
  quadratic, 55, 67–76
  for straight lines, 59–62
  trigonometric, 152, 159–64, 166–67, 171
equilateral triangles, 136
Euler's constant, 115
even-powered polynomial functions, 77
expanding, logarithmic functions, 111
exponential functions, 103–8
  applications, 119–26
  converting to log form, 109–10
  definition of, 104
  self-tests, 108, 122–23, 125–26
  solving of equations, 105–8
exponents, 1, 5–10
  base of, 5, 6, 9, 106–8, 117
  basic laws, 5–9
  coefficients and, 6
  constants and, 6
  fractional, 10
  fractions and, 7, 8–9
  logarithmic functions, 112–13
  in polynomials, 10, 11, 14
  positive and negative, 8, 9, 10
  pretest, 2
  raising a power to a power, 6
  self-tests, 7, 10
  zero as, 8
  See also rational exponents

factoring, 1, 17–24
  binomials, 18, 20
  by grouping, 19–20

difference of two squares, 20–21
  greatest common factor, 18–19, 21, 22
  negative terms, 20
  pretest, 2
  self-tests, 19, 20, 21, 23, 24
  sum/difference of two cubes, 23–24
  trinomials, 21–23
factors, bases as, 5
first-degree equations, 62
formulas, geometric
  area, 25–27, 191
  circumference, 26, 27
  distance, 24–25, 193
  midpoint, 24–25
  perimeter, 25–27
formulas, trigonometric, 164–77
  double-angle, 169–73
  half-angle, 173, 174
  Heron's, 192–93
  power-reducing, 173, 174
  product-to-sum, 175
  self-tests, 167, 172, 176
  sum and difference, 164–69
  sum-to-product, 175–76
fractions
  division of polynomials, 13–15
  exponents and, 7, 8–9
  improper, 98–99, 205
  partial fraction decomposition, 204–9
  reciprocals, 9
functions, 33–51
  composite, 33, 43–45, 49–50
  definition of, 33–34
  dependent variables, 34–35
  domain. *See* domain of functions
  exponential, 103–8, 109–10, 119–26
  general form, 75, 82
  graphs of, 35, 39, 45, 46–48, 52–102
  horizontal-line test, 46–47
  independent variables, 34–35
  inverse, 33, 45–51, 109, 127, 142–45
  logarithmic, 103, 109–26
  mapping, 34–36, 45, 46
  notation of, 34–36
  one-to-one, 46–48, 50
  operations on, 33, 41–43
  range. *See* range of functions
  self-tests, 36, 39, 42–43, 44, 48, 50
  standard form, 75, 76, 82, 105

trigonometric, 127, 134–45
vertical-line test, 35, 46, 47

G.C.F. *See* greatest common factor
general form, of functions, 75, 82
geometry. *See* basic geometry; formulas, geometric
graphics calculators, 52, 78, 82, 89, 140
graphs, linear inequalities, 196–204
graphs of functions, 39, 45, 46–48, 52–102
   amplitude, 141
   asymptotes, 82–102
   exponential, 104–5
   horizontal-line test, 46–47
   intercepts. *See* intercepts
   inverse trigonometric, 143
   linear equations, 62–67, 196–204
   logarithmic, 109
   period, 141
   polynomial, degree 3 or higher, 76–82
   quadratic equations, 67–76
   self-tests, 55, 58, 61, 65, 69, 75, 80, 87, 93, 100
   shifting technique, 71–76, 78–80, 82
   slope of straight lines, 57–59, 60, 61, 63
   straight-line equations, 59–62
   trigonometric, 127, 140–42
   vertical-line test, 35, 46, 47
greatest common factor, 18–19, 21, 22, 54
grouping, in factoring, 19–20
growth, exponential, 123–24

half-angle formulas, trigonometric, 173, 174
half-life, 124
Heron's formula, 192–93
horizontal asymptotes, 84, 85–86, 88–92
horizontal lines, slope of, 58, 60
horizontal-line test, 46–47
hypotenuse, 26, 132, 134

identities, trigonometric, 152–59
   self-tests, 155, 158
   simplification of expressions, 152–55
   verification of, 156–58
improper fractions, 98–99, 205

independent variables, 34–35
infinity, 83
initial side, of angles, 128
inner functions, 43–44
integers, 10
intercepts
   exponential functions, 105
   identification of, 53–57
   linear equations, 62–65, 197, 198
   polynomial functions, degree 3 and higher, 77–80
   quadratic functions, 55, 67, 68–69, 72
   rational functions, 88–92
   straight line equations and, 59–60
interval notation, 38
inverse of functions, 33, 45–51, 109
   trigonometric, 127, 142–45

k. *See* constant of proportionality

leading coefficients
   in horizontal asymptotes, 85, 88
   positive and negative, 77
linear functions, graphs of, 62–67, 196–204
linear inequalities, 196–204
   self-test, 202
linear programming, 200–204
lines. *See* asymptotes; horizontal lines; straight lines; vertical lines
line segments, 62
ln. *See* natural log
logarithmic functions, 103, 109–26
   applications, 119–26
   converting to exponential form, 109–10
   definition of, 109
   division property, 112, 113
   multiplication property, 111–12, 113
   natural log, 115, 117
   power property, 112–13
   self-tests, 110, 111, 113, 114, 116, 118, 122–23, 125–26
   solving of equations, 114–19

mapping of functions, 34–36, 45, 46
midpoint formula, basic geometry, 24–25
monomials, division by, 13

multiplication
  binomials, 12, 21–22
  commutative property, 21
  exponents, 5, 6–7
  of functions, 41–42
  polynomials, 11–12, 17, 20–21
multiplication property, logarithmic functions, 111–12, 113

natural log, 115, 117
negative angles, 128
negative exponents, 8, 9, 10
negative infinity, 84
negative leading coefficients, 77
negatives
  double, 15
  factoring of, 20
  quadratics, 67
negative signs, in polynomials, 12
negative slope, 58, 63
nonalgebraic functions. *See* exponential functions; logarithmic functions
notation
  of functions, 34–36
  interval, 38
  symbols, 57, 111
  triangles, 132
numerators
  asymptotes and, 85–86, 88, 90, 92
  exponents and, 7, 9
  improper fractions, 98, 205
  rational exponents and, 16

oblique asymptotes, 98–102
oblique triangles
  area of, 191, 192–94
  law of cosines, 187–91
  law of sines, 178–86, 188
obtuse angles, 179, 182
odd-powered polynomial functions, 77
one-to-one functions, 46–48, 50
origin of axes, 39, 65
outer functions, 43–44

parabolas
  axis of symmetry, 67, 68, 74
  constants and, 67, 72–74
  open up vs. open down, 67–68
  vertex of, 67, 68–69
parallelograms, perimeter and area formulas, 25, 27
parentheses, in notation, 38
partial fraction decomposition, 204–9
  self-test, 207
pentagons, 191
perimeter formulas, basic geometry, 25–27
period of curves, 141
polynomials, 1, 10–16
  addition, 11
  coefficients, 11, 77
  combining like terms, 11
  definition of, 10
  descending order, 14
  division, 13–15
  even-powered functions, 77
  exponents, 10, 11, 14
  factoring, 17–24
  graphs of functions, degree 3 or higher, 76–82
  linear equations, 62
  multiplication, 11–12, 17, 20–21
  negative signs in, 12
  odd-powered functions, 77
  pretest, 2
  rational functions, 82–83
  self-tests, 11, 12, 13–14, 15
  subtraction, 11
  variables, 10, 11
positive angles, 128
positive exponents, 8, 9
positive leading coefficients, 77
positive slope, 58
power. *See* exponents
power property, logarithmic functions, 112–13
power-reducing formulas, trigonometric, 173, 174
pretest, basic mathematical concepts, 1–5
product-to-sum formulas, trigonometric, 175
Pythagorean theorem, 26, 133, 147, 178

quadrants, standard unit circle, 127, 134–35

quadratic equations
  graphs of functions, 67–76
  intercepts, 55, 67, 68–69, 72
quadratic formula, 55

radians, 129–30
radicals, 1, 16–17
  pretest, 2
  self-test, 17
radioactive decay, 103
radius of circles, 26, 27
range of functions, 33, 39, 52
  acceptable members, 37–39
  asymptotes, 84
  definition of, 34–35
  exponential functions, 105
  inverse functions, 45–46, 109
  inverse trigonometric, 143
  trigonometric, 140–42
rational exponents, 1, 16–17
  pretest, 2
  self-test, 17
rational functions
  definition of, 82
  form of, 82–83
  horizontal asymptotes, 85–86, 88–92
  oblique asymptotes, 98–102
  partial fraction decomposition, 204–9
  steps for graphing, 88
  vertical asymptotes, 84–85, 88–92
rationalizing of denominators, 154
reciprocal fractions, 9
rectangles, perimeter and area formulas, 25, 26
reference angles, 134–39
relations, 34, 35
right triangles
  applications, 127, 148–50
  hypotenuse, 26, 132, 134
  Pythagorean theorem, 26, 133, 147, 178
  self-tests, 133, 148, 150
  solving of, 127, 146–48
  standard unit circle and, 134–36
  trigonometric functions, 127, 132–34
root of the function. See x-intercept
roots. See radicals

secant
  graphs of functions, 141–42
  identities, 153
  right triangles, 132–33
second-degree equations, 67
shifting technique, graphing and, 71–76, 78–80, 82
simplification, trigonometric expressions, 152–55, 169
sine
  angles, 134–39
  double-angle formulas, 169
  graphs of functions, 140–41
  half-angle formulas, 173
  identities, 153
  inverse functions, 143
  law of sines, 178–86, 188
  power-reducing formulas, 173
  product-to-sum formulas, 175
  right triangles, 132–33
  sum and difference formulas, 164
  sum-to-product formulas, 175–76
slope of lines, 57–59, 60, 61, 63
square roots, 16–17, 21
squares, perimeter and area formulas, 25, 26
standard form of functions, 75, 76, 82, 105
standard unit circle, 127–28, 130
  right triangles and, 134–36
straight lines
  equations for, 59–62
  linear equations, 62–65
  slope of, 57–59, 60, 61, 63
subtraction
  of functions, 41–42
  of polynomials, 11
sum and difference formulas, trigonometric, 164–69
sum/difference of two cubes, 23–24
sum-to-product formulas, trigonometric, 175–76
symbols, in notation, 57, 111
systems of inequalities, 196–204
  self-test, 202

tangent
  angles, 134–38

tangent *(continued)*
  double-angle formulas, 169
  graphs of functions, 141–42
  half-angle formulas, 173
  identities, 153
  inverse functions, 144
  power-reducing formulas, 173
  right triangles, 132–33
  sum and difference formulas, 164
terminal side, of angles, 128
test points, 196–98
TI (Texas Instrument)-89 graphics calculator, 52, 78
triangles
  area of, 25, 27, 178, 191–95
  equilateral, 136
  law of cosines, 187–91
  law of sines, 178–86, 188
  oblique, 178–94
  perimeter and area formulas, 25, 27, 191
  self-tests, 185, 189, 194
  trigonometric functions, 132–34, 136
  See also right triangles
trigonometry
  angle functions, 134–40
  angle measurement, 127–31
  applications, 146–51
  area of triangles, 178, 191–95
  basic formulas, 152
  double-angle formulas, 169–73
  graphs of functions, 140–42
  half-angle formulas, 173, 174
  Heron's formula, 192–93
  identities, 152–59
  inverse functions, 127, 142–45
  law of cosines, 187–91
  law of sines, 178–86, 188
  oblique triangles, 178–94
  power-reducing formulas, 173, 174
  product-to-sum formulas, 175
  right triangles, 127, 132–34
  self-tests, 131, 133, 139, 145, 148, 150, 155, 158, 163, 167, 172, 176, 185, 189, 194

  solving of equations, 152, 159–64, 166–67, 171
  sum and difference formulas, 164–69
  sum-to-product formulas, 175–76
trinomials
  factoring, 21–23
  sum/difference of two cubes, 23–24

unit circle. *See* standard unit circle

variables
  dependent, 34–35
  exponential equations, 104
  independent, 34–35
  polynomials and, 10, 11
  rational exponents and radicals, 16–17
  trigonometric equations, 159
vertex of parabola, 67, 68–69
vertical asymptotes, 83, 84–85, 88–92
vertical lines, slope of, 58, 61
vertical-line test, 35, 46, 47
vertices of triangles, 193

word problems. *See* applications

x axis, 53
x-intercepts
  identification of, 53–55
  linear functions, 62–64
  polynomial functions, degree 3 and higher, 77–80
  rational functions, 88–92

y axis, 53
y-intercepts
  identification of, 53–55
  linear functions, 62–64
  polynomial functions, degree 3 and higher, 77–80
  rational functions, 88–92
  straight line equations and, 59–60

zero, exponents and, 8

Printed in the USA
CPSIA information can be obtained
at www.ICGtesting.com
JSHW051654100624
64549JS00005B/174